CAD Systems in Mechanical and Production Engineering

Peter Ingham

Heinemann Newnes

To Clarice

Heinemann Newnes
An imprint of Heinemann Professional Publishing Ltd
Halley Court, Jordan Hill, Oxford OX2 8EJ

OXFORD   LONDON   MELBOURNE   AUCKLAND   SINGAPORE
IBADAN   NAIROBI   GABORONE   KINGSTON

First published 1989

© Peter Ingham 1989

British Library Cataloguing in Publication Data
Ingham, P. C.
    CAD systems in mechanical and production engineering.
    1. Products. Design & manufacture. Applications of computer systems
    I. Title
    670.42'7
ISBN 0 434 90870 3

Printed and bound in Great Britain by
Redwood Burn Limited
Trowbridge, Wiltshire

CAD Systems in Mechanical and Production Engineering

Contents

Preface

In the modern industrial environment it is unwise to consider any computer system, least of all computer-aided design, in isolation. Because of the current increasing emphasis on system integration, the boundaries between engineering subsystems are becoming more and more blurred. CAD is not just a matter for the drawing office – it influences all the other areas of the enterprise.

CAD is now an important tool in all branches of engineering – construction, electrical, chemical, electronic – but it is in mechanical engineering that is found the richest variety of applications. This book describes the various elements that make up the CAD function and explains how they fit together and interact with other parts of the computer integrated system, particularly those that are connected with production.

The book is intended to have a wide scope of application. It covers the subject matter of the standard BTEC unit CAD V, but will also be useful support for other courses, such as BTEC's CAE IV and CAM V. It should also provide a general background for the CAD widely taught in general mechanical and production engineering degree courses.

Chapter 1 broadly describes the development of computer systems in the engineering enterprise and shows how CAD fits into the integrated business system.

Chapter 2 deals with the hardware used in CAD – but from the user's point of view, the treatment deliberately being kept at a fairly low technical level.

Chapter 3 is a description of the more important pieces of software that make up a full CAD system. Since draughting systems are by far the most common packages used in industry, they are dealt with in some detail, but three-dimensional modellers, finite element analysis packages and database systems are also covered.

Chapter 4 develops some of the basic concepts of computer graphics. The techniques inescapably involve mathematics but the treatment is restrained and needs little beyond simple matrix arithmetic.

Chapter 5 concerns managerial aspects such as the selection of a system suitable for a particular application. It also explains the factors involved in the installation and running of a successful system.

Chapter 6 details some of the measures that have been taken to standardize CAD systems. Included are standards for passing drawing information from one system to another, standards for simplifying the writing of graphics programs and standards for general system intercommunication.

Included in most chapters are small scenarios which are intended to alert readers to the problems involved in various aspects of CAD. This technique is perhaps unusual in an engineering text but it has been found to be successful and popular with a variety of classes at different levels.

Most courses in CAD contain a high proportion of practical work on specific systems. This is beneficial but there is a danger that the students may become accustomed to a system that they are unlikely to meet in their jobs – for instance, at any given time there are over a hundred draughting systems on the market. Although the section on software *is* based on particular systems, these have been chosen carefully to illustrate the main features of *most* systems.

Also, because of this high emphasis on practical content, teachers have no difficulty in providing assignments based on the systems that they are using. This is one of the joys in teaching CAD. But expertise in CAD, as in all of the newer technologies, involves more than merely becoming proficient at using systems. Becoming an expert is a much easier matter than remaining an expert. Software and hardware are both developing at breakneck speed and it is vital for the future that our engineers should be equipped to cope with rapidly changing systems and applications. At the end of each chapter are a few ideas for assignments of a general nature which are designed to encourage students to discover information for themselves by reading some of the excellent specialized journals now available.

Most books on new technologies include a section on their social effects. Rather than including the usual balance sheet of credits and debits, a hysterical attack on CAD has been added as a postscript. If the book has been read with care, most of the arguments can be demolished with little effort.

Peter Ingham

Acknowledgements

I should like to thank Autodesk Ltd (AutoCAD), Deltacam Systems (DUCT, DUCTdraft) and Pafec Ltd (DOGS, DOGS3D, PAFEC 75, BOXER) for allowing me to use their excellent systems as examples.

1 *Introduction*

'Easy as ABC'

Scenario

Pinchbeck Products is an old-established firm which uses mainly traditional means to make a range of kitchenware. Mr Titus Oates, the Training Officer, is holding one of his periodic series of interviews with the trainee engineers. First on the list is Thomas Edison:

TO: Sit down and make yourself at home, Thomas. You have been with us now for...[*consults file*]...about eighteen months. I see that you are currently in the toolroom. Are they looking after you? Plenty of benchwork, I suppose. [*Goes into long and irrelevant account of his apprenticeship.*] Also, I see that you are taking a course at Borchester Tech. Are things going well?

TE: A lot of the course is very interesting. I am doing SPC at the moment.

TO: [*Looks at file again.*] I thought that you were doing a technician course.

TE: I am, as a matter of fact. SPC means 'statistical process control'.

TO: Of course it does. I was just testing you. Well done, young man. What else are you doing?...

Later that morning, Mr Oates phones the Chief Draughtsman:

CD: Charles Darwin speaking.

TO: Titus here. What are you doing for lunch? I have one or two things that I want to discuss with you.

CD: OK then. See you in 'The Zoom and Pan' at 12.30.

Settled over lunch:

TO: The trouble is that there seems to be a new three-letter word
          out every week - CAD, CAM, CAE, CIM. Where will it end? I
          know what each of the words 'computer', 'aided' and
          'design' means. But when they are put together, what on
          earth do they mean then? What's it all about?

CD: Well, CAD is an easy one. It means using computers
          interactively to help with the design process.

TO: What's new about that? We were using computers in design
          in the sixties.

CD: Yes, but in those days, you didn't use them interactively.
          You wrote a program, handed it into the Computer
          Department with the data, and had no more contact with it at
          all until the answers were produced.

TO: Or a list of errors in my case.

CD: Nowadays, the designer sits at a terminal and works with the
          program while it runs. The computer and the human work
          together as a team, each partner doing what they are best at.

TO: But doesn't it just mean draughting with the aid of a
          computer?

CD: Many people think so. But CAD is far more than draughting.
          It includes many more design functions than draughting.
          There are systems which will display realistically the
          appearance of a part, systems which will analyse the strength
          and performance of a component and so on.

TO: That sounds useful. I know we don't use CAD yet, but we do
          have CAM...don't we? I've seen all those machine tools on
          the shopfloor, driven by punched tape.

CD: That's just numerical control. CAM is similar to CAD - all the
          manufacturing operations are assisted by computer: process
          planning, numerical control, scheduling and all the other
          functions.

TO: And I suppose CADCAM (or is it CAD/CAM?) is using both of them?

CD: You've got it. But both are linked together.

TO: How about CAE, CIM, MRP, MRP2 ... ?...

Later in the afternoon, Mr Oates is still interviewing the apprentices and is in the middle of his talk with Tilly Eulenspiegel:

TO: So you are taking CAD, are you? Funny how a lot of people still think that it is just draughting. Of course, it includes...

■　　■　　■

To explain some of the terms which have been baffling Pinchbeck Products' Training Officer, it will be helpful to trace the development of modern engineering systems over the last few years. Despite the sudden emergence of computer systems, this development has been a steady process. In this, as in many continuous processes, it is difficult to take a snapshot of conditions at a single instant and give it a label. When artists name a colour of the spectrum 'red', they do not normally mean 'light with a wavelength of 550 nanometres'; they mean something that is less exact - the quality of 'redness'. Similarly, in engineering systems the term CIM, for instance, refers to a tendency rather than to one precise type of system. This is the reason why many published definitions of the three-letter acronyms commonly used today seem so broad (and sometimes conflicting). Some of the terms that were bothering Pinchbeck Products' Training Officer may be defined as follows:

CAD Computer-aided design: a set of tools for automating certain of the steps in the design process.

CAM Computer-aided manufacture: the application of computers to the manufacturing process.

CAE Computer-aided engineering: the concept of automating all the steps involved in creating a product and enabling all engineering disciplines to share the data created by others.

CADCAM Computer-aided design and manufacture: this was once used to describe computer-aided design, draughting and manufacture using interactive workstations but is now often used to encompass the whole area of the computerization of manufacture.

CIM   Computer-integrated manufacture: the concept of
                integrating all the functions of the design and manufacture
                of a product into a unified system using computers.

You may have a little difficulty in distinguishing some of these. Don't
worry about it. They are all taken from reputable sources; the wording has
been modified a little but the sense retained. It would be possible to collect
many more definitions which would be equally vague and some of which
would be contradictory.

System Development

Departmentalization

Early engineers, we are reliably informed, were far superior to us. They
could invent new products, design them in detail and even make them, if
required to do so. They knew everything. Of course, this was because
there was not much to know at the time. There were few materials in
common use, there were few manufacturing processes and products
were primitive compared with those of today. Because competition was
limited, they could get away with gross over-design which was often
necessary because of the lack of adequate analytic tools. Production was
often a hit-or-miss business, involving a good deal of selective assembly.
Their attitude is summed up by the remark supposedly made by
I.K.Brunel to the directors of the Great Western Railway: 'You should be
glad that that bridge fell down: I was planning to build thirteen more to
the same design.'

   Gradually, new materials and processes were introduced, technologi-
cal advance resulted in a higher degree of product complexity and
increased competition forced designs to be more cost-effective. With
these developments arose a need for specialists such as materials
scientists (because of the new materials and processes) and stress
engineers (because of the leaner and more closely specified designs).
Similarly, with the parallel advances in production and the demand for
higher accuracy and repeatability with large production volumes, a need
was created for production engineering specialists. As the variety of
specialisms and the number of specialists increased, the engineering
function became departmentalized.

   Departmentalization was inevitable in industrial firms as an admin-
istrative convenience and as a means of localizing expertise, but it has
attendant problems. The most apparent of these are (a) the difficulty of
retrieving good quality information and (b) the difficulty of controlling
departments effectively.

   If you prepare a chart showing just the major departments of a firm
with which you are familiar, and fill in connecting lines showing

communication between them, you will probably be surprised by the 'plate of spaghetti' that results - Figure 1.1 shows a comparatively simple case. The information flows from one centre to another in various ways - telephone calls, forms, verbal messages carried by progress chasers and many more. It is a marvel that large firms can function at all with such a haphazard system. Because of the informality of communication, data is often duplicated and the different versions existing at one time are sometimes inconsistent. With such casually designed communication channels, it is difficult to guarantee the quality of the information that is essential for the smooth and profitable operation of the business.

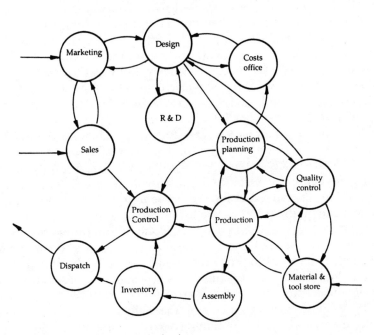

Figure 1.1   *Departmental intercommunication*

A connected problem is one of difficulty of control. It is not unknown for managers to make correct decisions intuitively but most would prefer to base them on good quality information. Even if information is consistent and up to date, it is difficult, in a highly departmentalized firm, to gather data from the various centres rapidly and conveniently enough. Managers need historical data to help them to plan future operations and they need rapid access to current data so that they can monitor and control progress. If information is dispersed around the firm in filing cabinets, in in-trays and in transit, then obtaining reliable data quickly involves a considerable effort.

   As well as the lines of data flow, we can also draw lines of control. These represent the management commands that drive the system, ensuring that the correct process is performed at the correct time. In many cases, these are the same lines as those of information flow and the same problems exist. Control commands in a firm can become distorted when passed along a chain (and often are).

Enter the Computer!

Computers were introduced in the 1950s but were used rather differently from the way they are used now. They were mainly engaged in calculation in 'batch mode'. Batch mode involves submitting a program and data to a computer, going away, and waiting until the results are produced. The user has no contact with the program as it runs. This is rather like writing a letter and posting it, then waiting for a reply. If we have unfairly insulted someone in a letter, there is no way in which we can amend the message while it is in transit. Batch mode is fine for computers; it uses them very effectively and was the ideal way to work when they were rare and costly machines. However, it does not use the human resource efficiently. Computers used like this did not improve, even in the slightest way, information gathering or control.

As computers became cheaper and more common, the need to use them efficiently at the expense of human time and effort decreased. Many firms in the 1970s entered the fields of CAD and CAM. Specialized packages became available in bewildering numbers. Often (and draughting systems are an example), they were used 'interactively': many users shared the program and controlled its progress by means of a keyboard. Interactive mode is like making a telephone call. In this case, we *are* in contact with the message. This interactive use of systems is now the general rule, except for applications such as finite element analysis that need a lot of 'number-crunching'. Using systems interactively is usually a waste of computer power but is an effective use of the human resource.

Typically, a firm would enter the field in one area - possibly in computer-aided part programming or computer-aided design. The smart firms soon realized that it was possible to pass information from a draughting system to a part-programming system by means of a file. This was the first sign that computers might help with the communication and control difficulties outlined previously. Apart from this, adding 'CA' in front of any of the traditional engineering functions is of little organizational significance and the underlying structure, with its attendant problems, remains exactly the same as the uncomputerized version.

A typical firm, at this time, had invested a fair amount of money in computer systems, numerically-controlled machine tools were becoming more common and robots were seen occasionally, but there was little communication between these 'islands of automation'.

The next stage in the evolution of the integrated system is to join all the subsystems into one large system. There are obvious links that can be

made easily - we have mentioned the one between CAD and computerized NC part-programming. However, the desirable links between
the subsystems are many and varied; they cannot be achieved without a
lot of time and effort. Since most systems have been developed bit by bit
over a fair period of time, it is common, especially in smaller firms, for the
subsystems to have been bought from different sources. If linking is
achieved by translating from one format to another, then the resulting
system is not much better than the system shown in Figure 1.1.

True integration can be attained only by using a 'database'. Databases
will be described in some detail in Chapter 3; for the moment they may be
assumed to be just centralized stores of information. Figure 1.2 shows a
simplified version of such an integrated system. Communication
between all the component subsystems can now be achieved through this
common source of information. Ideally, the data is non-redundant: only
one copy of any piece of information is kept. Therefore there are not the
conflicting versions that exist in the traditional departmentalized organization. In practice, this ideal is hard to attain. For instance, it is possible to
set up various versions of the geometry of a part - a three-dimensional
model for calculation of mass properties and collision detection; a finite
element mesh for calculation of stresses and deflections; a traditional
orthogonally projected drawing as a basic record. For practical reasons,
the different representations are often held side by side in the database.
The difficulty then is to ensure what is termed 'associativity' - to link the
different models together so that a change in one results in a corresponding change in the others.

Figure 1.2   *Communication through database*

Another property of the true database is that of data independence. Because of the lack of redundancy in the data, there should be some mechanism to ensure that each system can view the data in a way that is appropriate to its own needs. Again, most current systems do not permit this. What in many applications is termed a 'database' is often no more than a large file of data.

Despite these shortcomings, the system shown in Figure 1.2 has many advantages. It is the first of the systems that we have described that can be truly called 'integrated'.

Integration

In the application of computers to the manufacturing function, technical progress is very fast. Currently, there is the capability to computerize all the individual functions that make up the 'product cycle' from demand for a product to its production. In most firms, progress towards a fully computerized system has been rather nervous but is accelerating. Of course, 'design and manufacture' is only part of the requirement. The ultimate aim is the 'integrated business system' where finance and marketing are also combined into the system. It is probable that most firms will never reach this stage fully but even a partial implementation should pay high dividends in profitability.

Up to this point, we have been discussing the mechanization of the more clerical tasks in the product cycle - the design of the product, the NC part-programming, operation sequencing, ordering of raw materials and other manufacturing operations. On the shopfloor, computerization of production processes has been developing from the mid-1950s in the US and a little later in the UK. The development has normally resulted in further 'islands of automation', for instance, FMS cells which are computer-controlled but are independent of any other centre in the plant. It has long been recognized that it would be beneficial if these could in some way be connected to a centralized system so that they could be more efficiently controlled. Here again, the communication problems have been severe and, because of the widely differing standards of communication interfaces and formats, to integrate these isolated centres into a centralized system has needed high investment of time and effort. There is now a move to standardize means of communication by MAP and TOP; these will be discussed later.

We have now arrived at a fully computer-integrated system. It is unlikely that many firms will develop a full integrated business system, but most successful ones should go a long way towards doing so.

Another advance, which is little to do with integration but which will affect engineering users profoundly, is the incorporation of artificial intelligence into user interfaces. A considerable amount of research is being carried out into 'expert systems'. These are systems which permit the expertise of a specialist to become available to ordinary non-specialized users. Their operation has been described as that of 'cloning experts'. It seems likely that the use of these in some of the tasks carried out in engineering will become commonplace before the end of the 1990s. At the beginning of the decade they are only suitable for small, restricted areas of expertise - an example in design is bearing selection. However, it may well be that with the development of the newer generation of computers, their application will be extended and that large areas of expertise will be held in 'knowledge bases'.

It is interesting to speculate what the role of the engineer would be in a full integrated business system with expert system interfaces. It seems apparent that there would be a high demand for broadly-trained users who could use the system in an effective way. In an extreme case, one skilled user could perform all the functions that are now done by specialist departments. We might even be forced to become the all-round engineers that were discussed at the beginning of this chapter.

Problems

1. (In this and similar exercises, you are recommended to write definitions in your own words in a small notebook.) Write down brief definitions of the following acronyms:
  (a) CAD
  (b) CAM
  (c) CAE
  (d) CAPP
  (e) CADCAM
  (f) CIM
  (g) CAPM
  (h) MRP
  (i) MRP2
  (j) AI

2. How does work flow through the Drawing Office of some firm with which you are familiar? Explain:
  (a) where jobs come from;
  (b) what the procedure is for distributing them;
  (c) where designers get information from (e.g. materials specifications);
  (d) how the DO links other departments, e.g. Costs Office.

3. We stated that few firms will develop a fully integrated system but most successful firms will go a long way towards it. Do you think that these firms will be necessarily large? Are computer-integrated systems more suitable for big firms or for small/medium ones?
  Give reasons for your opinions.

2 Hardware

Introduction

'What do you mean, "It's driven by a mouse"?'

Scenario

We return to Pinchbeck Products. The Managing Director has invited two sellers of CADCAM systems to show him their products. Also present are the Engineering Director, the Chief Designer, the Computer Manager and the Chief Accountant. The first salesperson is from a firm which markets a microcomputer-based system called GrotCAD:

SP: Good morning, ladies and gentlemen. I should like to introduce you to GrotCAD, the latest in draughting systems - minicomputer performance for microprice. Later, I shall be talking about GrotCAM, the CAM system which interfaces with it. If any questions occur to you during my presentation, do not hesitate to interrupt. The system is based on a variety of 16-bit micros, but we recommend the Pomegranate PC. Each workstation needs a co-processor, 640k of RAM and a Procrustes Graphics Card. Input is via a QWERTY keyboard and the cursor is driven by a mouse…

CA: How much does all that cost?

SP: Well, believe it or not, you will have change out of £5000 per workstation for the hardware. The software is also costed very competitively. We charge £5000 for the first seat but each subsequent station costs just £500. But let me get on with discussing the performance. We call it 'The electronic drawing-board and tee-square'. It can do anything that can be done by traditional means but far more accurately, and often faster…

MD: You said that the cursor is driven by a mouse. Could you possibly amplify that. A little.

SP: [*Smiles appreciatively.*] I expect you are used to seeing larger systems where input is from puck or stylus acting on a bit-pad. Well, we can offer those options also.

MD: Thank you.

CM: Can the system be networked?

SP: We are working on that at this moment in time. Our systems people should come up with a solution within the next month or so. However, I don't think that you need bother about networking. Each workstation has its own local storage - a hard disk holding a massive twenty megabytes of store...

A little later, another salesperson is giving the same group a presentation:

SP: Good morning, ladies and gentlemen. I should like to introduce you to PlutoCAD, the latest in CADCAM systems, mainframe performance for miniprices. We call it 'The electronic engineer'. If any questions occur to you during my presentation, do not hesitate to interrupt. The system is based on a MAX mini supporting a maximum of twelve users...

CA: How much does it cost?

SP: This, of course, depends on the configuration. As a ballpark figure, I should guess that you would have change out of £60 000 for four workstations. Let me tell you about what PlutoCAD can do for you...

MD: Can it be driven by a mouse?

SP: [*Smiles appreciatively.*] I see what you mean. The system is user-ingratiating and almost anyone could use it with a minimum of training.

MD: Thank you.

SP: To get back to the main thrust of my presentation...

Next day the group meets to discuss their findings:

MD: Well, I don't mind telling you, I was pretty punch-drunk after those presentations yesterday. Has anyone reached a conclusion?

CA: As far as I'm concerned, GrotCAD has very clear advantages. If we start with six workstations, the cost will be less than £40 000. That would seem to be a satisfactory figure.

ED: But that would just give us six separate systems. There would be little advantage over the existing manual methods. I would suggest that we plump for PlutoCAD - £100 000 for six workstations, say. We would also need a good plotter to give customers the right impression. That would cost us another £20 000 or possibly a bit less.

CD: I would support that - we could do with eight workstations, though.

CA: I don't want to appear obstructive but I think it would be best to try the water with a toe before we take the plunge.

CM: I'm not so sure that either is the correct solution. We have a perfectly good existing computer system. Why don't we enhance the existing processor, add a little extra backing store and go for a completely centralized system. The clear advantage would be...
      . . .
      . . .
      . . .

MD: Well, thank you for that free exchange of completely different views. I think it might be useful if we set up a task group to look into the matter. It might pay us to get an outside consultant to advise them. I might even be able to find out what a mouse is then.

■   ■   ■

It is sometimes said that if a CAD system is well designed the users need know nothing of the hardware beyond the location of the on/off switch. It is true that system designers try to shield the ordinary user from the hardware and sometimes they are reasonably successful. However, there are types of user who need a reasonable knowledge of how the physical components of a system work. Initially, as we have just seen, members of staff will be involved in the selection of a system. If the selected system is to match the needs of the company, then it is helpful if they can have some idea of the operating characteristics of the system hardware so that conflicts of cost and technical requirements can be resolved. For instance, Pinchbeck Products' Chief Accountant may be concerned only with keeping the total outlay as low as possible; the Computer Manager may be interested only in the technical specification. The Chief Engineer, naturally enough, will be interested only in the system's ability to support an efficient draughting facility. Each, in isolation, would probably arrive at a different specification which might not be the best for the firm's interests. If all the interested parties had some knowledge of what facilities they would get for their money, then it is likely that, after some discussion, a compromise solution would be reached which would meet the requirements reasonably closely.

The ordinary day-to-day user needs to know very little about a system. However, it is not in the nature of engineers to use equipment without having a fair idea of how it works and most CAD users are keenly interested in the hardware components of a system. It might be argued that this improves their efficiency in using the system.

There is another category of user: the system builder. As we have discussed in the introductory chapter, there is a strong tendency for systems to become integrated. This often means that subsystems from more than one vendor have to be linked together. Staff engaged in this work need an intimate knowledge of all the elements of a system, both hardware and software.

The Elements of a CAD System

We shall start by looking at CAD systems simply as a set of 'black boxes' joined together by wires, as shown in Figure 2.1. The computer takes in data and programs, stores them temporarily and processes them to produce answers. Data and programs are entered initially by users but it would be intolerable if this had to be done each time the system was used. Typically, the system is 'transparent' to the user, who just wishes to turn on the power and use the system immediately. Backing (or auxiliary) store is therefore provided, and this is used to hold programs and data on a long-term basis; programs and data must be transferred to the computer's internal storage for processing.

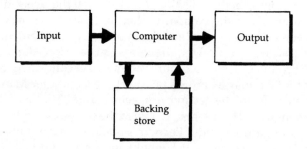

Figure 2.1 *A computer system*

Data, programs and commands are entered into the system through input devices. Answers and other forms of computer communication with the user are given through output devices. Input and output are often referred to collectively as 'I/O'.

Examples of each black box are:

Input device: a keyboard
Computer: an IBM PC
Backing store: a floppy disk
Output device: a dot-matrix printer

It is important that the common boundaries (or interfaces) of each connecting piece of equipment should be compatible. Although it is common to refer to the boundaries as 'interfaces', the same word is also used for the hardware provided for communication at each boundary. There are several interfaces which have been developed by manufacturers for special purposes and which have been used so widely that they are regarded as standard. Examples are the RS232 serial interface and the Centronics parallel interface. Another term often met is 'protocol', which means an agreed convention for the control of transmission of data.

It is also common to talk about the 'Human–machine Interface', which means the way that users communicate with the system and the way that the system communicates with users. It is critically important that this should be well designed or user speed will decrease and user errors will increase. Unfortunately, designing a good user interface has turned out to be a good deal harder than designing a hardware interface, because the principles are not so well understood. In spite of a lot of research into the human–machine interface, it is still common to hear users, especially inexperienced ones, complaining about the lack of help that they get from systems, the unintelligibility of manuals and the general difficulty of using CAD systems.

It might be claimed that users need know very little about the contents of our black boxes, just as car drivers need know nothing of the workings of the internal combustion engine. As we have previously argued, though, it helps to know something of the way the components of a CAD system work in order to compare competing systems and so the operation of the more common devices is described in a little detail.

Communication Between Devices

The arrowed lines joining the boxes in Figure 2.1 are usually cables along which data passes. Although the ordinary user need know nothing about the *technical* details of data transmission, it is useful to appreciate the way in which signals pass between computer and devices.

Data passes in the form of bit patterns and it is convenient to think of these patterns as codes for characters. The most widely used standard character code is the American Standard Code for Information Interchange or ASCII (pronounced 'Askey'). In this code, the characters that can be found on a normal typewriter (and some that cannot) are represented by a pattern of seven bits; for instance, 'A' is represented by '1000001' (decimal equivalent 65) and 'a' by '1100001' (decimal equivalent 97). The seven bits can hold 128 different codes which is easily enough for upper and lower case letters, numbers and a selection of other useful characters such as '*'. Some of the possible codes are reserved for special purposes - they represent not 'alphanumeric' but control characters.

Although each ASCII code takes seven bits, data is sent in eight-bit form. The extra bit is known as the 'parity' bit and can be used to detect some transmission errors. If the parity bit is a '1', to make the total number of 1s in the code word even, 'a' will be transmitted as '11100001'. This is 'even parity'. If a single bit were in error, either a 1 being sent as a 0 or a 0 as a 1, then the total number of 1s would be odd. This error would be detected by hardware and some action taken. There is also 'odd parity' checking, where 'a' would be sent as '01100001'. Some systems do not use parity check, while others use more than one parity bit so that not only can an error be detected, it can also be corrected.

If you look at the cables joining the devices on a computer system, you will notice that often two different types of cable are used. Some bits of hardware are connected by what look like the cables on domestic power leads, while others may be connected by flat ribbon cables. The round ones are often used for 'serial' transmission, the flat ones for 'parallel' transmission. In serial transmission, the separate bits of the message move along one wire in the form of a 'bit train', one after another. In parallel transmission, the bits travel side by side along eight wires (Figure 2.2). The transmission can be in a regular stream of characters -

'synchronous' transmission - or at irregular intervals - 'asynchronous' transmission. Synchronous transmission can occur when files are being transferred between the computer and backing store; it is most efficient in this case to send the data in long, continuous streams. Asynchronous transmission can occur between a keyboard and the computer since, even with a skilled typist, the intervals between characters are irregular in the timescale of the computer. In practice, neither serial nor parallel transmission is quite as simple as this. As well as the data bits, other signals are required for control purposes.

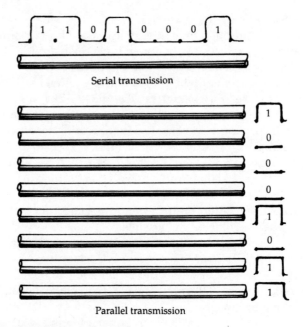

Serial transmission

Parallel transmission

Figure 2.2   *Types of data transmission*

The speed of transmission is measured in bits per second. A unit often used is the baud, which for binary transmission is one bit per second. There are several standard transmission rates, 110, 1200, 4800 and 9600 baud being common.

The type of transmission used depends on the application and it is normal to use more than one type in a system. For instance, if we were sending data across an ordinary telephone line, then it would be impracticable to send it along eight or more wires - serial transmission is appropriate. On the other hand, if we were sending output to a closely-located dot-matrix printer, it would be possible to send it in parallel along the eight wires of a ribbon cable.

Output devices

Monochrome Displays

Although colour displays are becoming common in engineering applications, many mechanical engineering systems still use monochromatic displays. Some of the many possible types are:

Refresh tubes
Direct view storage tubes
Raster scan tubes
Liquid crystal displays
Plasma panels

We shall discuss some of these in a little detail.

Refresh Tubes

Refresh tubes are based on the familiar cathode ray tube used in oscilloscopes (Figure 2.3). The face of a sealed glass tube is coated with a phosphor compound which glows for a while when it is hit by high-speed electrons. The picture is drawn by a beam of electrons which can be directed over the face of the tube and which can be turned on and off. The beam is directed to a desired point by a deflection yoke around the tube. The picture is drawn by the electron beam much as we write with a pen and so this type of display is sometimes called 'calligraphic'.

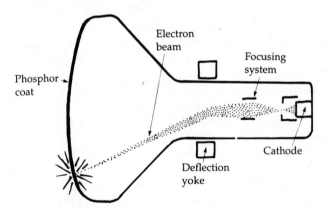

Figure 2.3   *Cathode ray tube*

Because of the properties of the phosphor coating, the picture fades very rapidly and it is necessary to redraw (or 'refresh') it as long as is needed to keep it on the screen. Humans display what is called 'visual retention', a property of the eye and brain that enables us to convert rapidly-changing separate images into continuous motion. In order to maintain the illusion that the drawing is continuously being displayed without 'flicker', the rate of refresh must be at least 25 times each second (or 25 hertz). Modern displays are refreshed at about 60 Hz and so are flickerless.

It would involve a lot of computing time to send plotting information 60 times a second to the display, so refresh screens are provided with storage to hold all the information necessary for drawing the picture. Since the display consists of lines, it is only necessary to hold the end points of straight lines, and the centres and radii of circles and similar data. The storage therefore can be quite compact. There are various names for this kind of storage, commonly called a 'display buffer'. Because of the short cycle time between screen refreshes, it is possible to draw only a limited number of lines on the screen, so pictures cannot be very complex nor can shapes be filled in easily. The technology permits limited colour facilities.

It would seem from the preceding remarks that refresh tubes have not much to recommend them. However, because of the rapid refresh rate, it is possible, by changing the display buffer, to change pictures more rapidly than with any other type of screen and so this kind of display is used for animation purposes, for instance when it is required to show vibrating structures.

Direct View Storage Tubes

A modification to the refresh tube avoids the need for image refresh and so avoids the restriction on the complexity of picture which can be drawn. In the direct view storage tube (DVST), the picture, once drawn on the screen, is maintained without refresh by a system of 'flood guns'. The picture can be very detailed indeed and this type of display is used for applications such as electronic CAD where complex circuits need to be displayed in a clear way.

The DVST has the advantage over refresh graphics for such purposes, but has its own attendant disadvantages. In order to make the slightest change, the picture must all be deleted first; this takes place with a characteristic flash. Because of this the DVST is clearly unsuitable for animation. Also, since the picture is not being continually redrawn on the screen, if it is unchanged for a long period, it burns its way permanently into the phosphor. DVSTs are fitted with an automatic dimming facility, which prevents this. Even the limited colour potential of the refresh tube is difficult to achieve on the DVST. The more expensive screens of this type are provided with a modest refresh capability, usually in a different colour from the storage display.

Raster-Scan Displays

A raster is a 'predetermined pattern of scanning lines providing uniform coverage of a display area' - an example being the picture on a domestic television set. In raster-scan displays the electron beam sweeps across the screen from side to side and top to bottom, being turned off and on at fixed intervals of time. The resulting picture is composed of a series of dots called 'picture elements' or 'pixels' (Figure 2.4). Because the scanning pattern is repeated constantly, the raster-scan display is a form of refresh tube, but is not calligraphic because lines are not drawn directly from point to point and pictures are more like brass-rubbings than pen drawings. To take advantage of the visual retention property of human perception, the whole screen is scanned at about 60 Hz. On some displays the scanning is 'interlaced', only alternate lines being drawn on consecutive scans, permitting a flicker-free display at half the real refresh rate. This technique was first used in motion pictures where each frame is shown twice at 24 Hz.

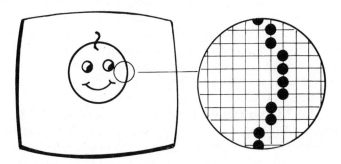

Figure 2.4   *Raster-scan display*

Continuous research into graphic displays has resulted in slow, but steady, increase in the number of pixels that can be displayed - the 'resolution'. At the time of writing the highest resolution on screens used for CAD is 2048 pixels across the screen by 2000 pixels up the screen. This gives a high quality picture but raster screens are still unable to achieve the detailed pictures displayable on DVSTs, the difficulty here being that the whole screen must be redrawn at each refresh cycle.

Since the raster-scan display has to be refreshed, it must be supplied with its own storage or 'frame buffer'. The difference between this and the refresh tube's display buffer is that the frame buffer must hold information about every displayable pixel, whereas the display buffer needs only information such as the end points of lines. Even monochrome screens need large frame buffers - a typical commercial screen of 1024 x 1024 pixels needs 131 072 bytes - and much more for colour screens.

The physical shape of the display depends often on the application. Office workers are used to looking at A4 sheets and so screens used for word processors are often higher than they are wide - the so-called 'portrait' aspect ratio. Engineers are used to drawing sheets and screens used for such applications as draughting are wider than they are high - the 'landscape' aspect ratio.

Since pictures are composed of pixels, most of the lines and all of the curves show a staircase effect or 'the jaggies'. It is possible to reduce this effect on colour screens using a clever device called 'anti-aliasing'; on monochrome screens, we can do nothing about it. The screen must be used as a sort of rough sketchpad and if we want clearer detail, then we must examine the picture locally at increased scale. Three measures of the detail can be achieved on a display: addressability, resolution and visibility; these can all be different.

Addressability.   Even though only, say, 1024×1024 pixels can be displayed on a screen, pixels can be addressed by software as though there were 4096×4096. In this case, each displayable pixel would correspond to 16 addressable ones. This permits the user to zoom into the display up to the limit of addressability by hardware, which is very much faster than having to recalculate pixel addresses by software.

Resolution.   This is the size of the actual pixel grid and on commercial screens is often 1024×1024.

Visibility.   In order to maintain the landscape aspect ratio of a screen used in engineering applications while keeping the scale constant in the horizontal and vertical directions, some of the pixels are not displayed. The aspect ratio is usually about 4:3 and a common visibility is 1024×784.

Raster technology is constantly improving and this type of display is the one most commonly found in CAD systems.

Colour Displays

The working of a colour display is much like that of a monochrome display, except that there are three electron guns and the face of the screen has three corresponding phosphors. The most common arrangement is that the three phosphors are deposited on the screen in triangular patterns of dots called 'triads' (Figure 2.5). Each of the three phosphors can be struck by its own electron beam and the beams are aligned to the phosphors by a perforated metal sheet called a 'shadow mask'. Each phosphor displays a primary colour and, when viewed from a distance, they appear as well-defined colours. This is possible because of another property of human perception called 'spatial integration'.

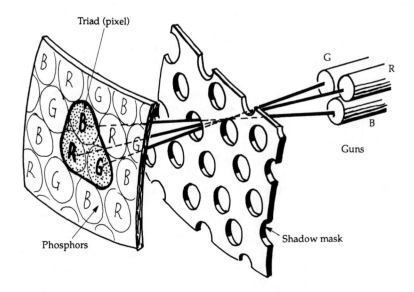

Figure 2.5   *Colour displays*

The three primaries can be selected in several ways, the most common being red, green and blue. This may seem strange - we are used to mixing paints where the three primaries are red, yellow and blue. Light, however, mixes differently from pigments: red and green make yellow, yellow and blue make cyan (turquoise) and red and blue make magenta. If we mix red, green and blue, we get white. This is called 'additive' mixing and it gives different results from the 'subtractive' mixing done with paints (Figure 2.6). Another common set of primaries is cyan, magenta and yellow.

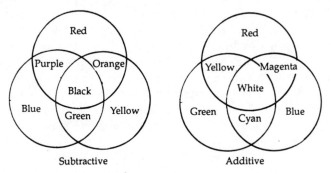

Figure 2.6   *Colour mixing*

Each of the three primaries can be displayed in a fixed number of intensities. If, for instance, each primary can be displayed in 16 different intensities, then the total number of mixtures available is $16 \times 16 \times 16$ or

4096, varying from black to white. The display would then be specified as having a 'palette' of 4096 colours. This would be a fairly cheap colour display. A higher quality (and more expensive) display might have 256 different intensities for each primary, giving a palette of 16.7 million colours.

All the palette of colours is not simultaneously displayable because of the size of the frame buffer which would be needed. A common colour display with a resolution of 640 × 480 pixels can display 16 colours selectable from a palette of 4096. Since 4 bits are needed to hold colour codes in the range 0-15, the frame buffer size is 640 × 480 × 4 bits or 153 600 bytes. If all the 4096 colours were displayable simultaneously, the frame buffer size would be increased by a factor of three, because 12 bits would be needed to hold the colour codes. In fact, 16 colours are usually adequate for applications where the colour is used just to make the information on the screen clearer - plant layout is an example. In applications such as the production of shaded, highlighted images we need more colours; 256 is a common number. A display which can show 16 colours simultaneously is known as a 4-bit plane screen; one which can show 256, an 8-bit plane. In order to reduce the size of frame buffers, some displays take advantage of the fact that colour usually occurs in blocks and use a more elaborate display storage system called 'run-length encoding'.

It is difficult for users to achieve a desired colour by specifying mixtures of different intensities of three primaries such as red, green and blue, and some manufacturers have their own method of specifying colours which is more easily used. A common one is the Munsell system used by Tektronix. This is based on three qualities of the desired colour - its *hue, light* and *saturation.* Hue is the basic colour - whether it is red or yellow or yellowish red. Light is the amount of white that is present in the colour. Saturation is the most difficult to appreciate without a demonstration on a screen - it is a measure of the brightness of the colour. Colours are visualized as lying in a double cone (Figure 2.7); hue is quoted as an angle - for instance, 120 degrees is red; light and saturation are quoted as percentages. After a little practice, it is possible to become quite proficient in the specification of colours using this method, far more so than if one were restricted to mixing primary intensities additively. Other systems are also used, but the H–L–S system is probably the most common.

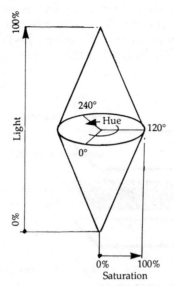

Figure 2.7   *H–L–S colour system*

Other Types of Display

The types of display that we have described have all been based on the cathode ray tube. This is still the most widely used technology in spite of its marked disadvantages: it is bulky and heavy, it uses a lot of energy, it is easily damaged and it uses dangerously high voltages. There is also some evidence, though rather inconclusive, that prolonged exposure to CRTs is hazardous to health.

Other types of display are widely used (for instance, on digital calculators and watches – liquid crystal and light-emitting diode displays) which do not have the disadvantages of the cathode ray tube but have others of their own. We shall describe briefly two types of display which are comparatively flat and so are suitable for use as screens on portable computers.

Liquid Crystal Displays (Figure 2.8)

Some compounds have the property of forming crystals that can flow like liquids at a certain temperature. A thin layer of one of these is sandwiched between two flat plates which are transparent polarizing layers with directions of polarization at right angles to each other. When an electric field acts at a particular area of the liquid crystal, it produces a dark section in that part. Displays are organized like the dots on a dot-matrix printer, being commonly in 5 × 8 blocks that can hold an alphanumeric character. When a pixel has a current passing in both horizontal and vertical directions, it shows darker than those which have either no current in either direction or a current just in one direction. This is called 'matrix addressing' and is used in several storage devices. LCDs have limited contrast and no colour facility but they require little power and can be made very flat.

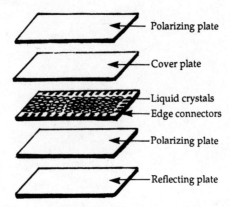

Figure 2.8   *Liquid crystal display*

Plasma Panels

Light is generated when a current is applied to an ionized inert gas; a neon tube is an example of this effect. A plasma panel is two transparent glass plates with a layer of inert gas (or plasma) sandwiched into the thin gap between them. Edge connectors are used to matrix address regions of the plate in a similar way to the LCD. Plasma panels, however, give very high contrast displays: resolutions high enough for graphics applications have been developed. They are currently monochrome, but are very flat and are unique among displays in being transparent so that photographs can be back-projected on to them.

Other Displays

Considerable money is being spent on the development of screens that can display three-dimensional images. Several of these have been announced; one uses a vibrating mirror and another displays two plane-polarized pictures alternately. The first does not need any special viewing equipment, but gives a rather blurred image, but one that is certainly three-dimensional. The latter gives quite staggering three-dimensional realism, but requires the user to wear special glasses. Both are very expensive.

Plotters

The users of a CAD system draw on a graphics screen and the resulting drawing can be kept on backing store. Although many firms use backing storage as a drawing store and keep readable versions of drawings on microfilm, invariably paper drawings are still needed. Even when a computerized drawing storage system is used, there is still a need for drawings on paper for discussion or for sending to customers. Many CAD users, due to mistrust or other reasons, retain paper drawings in a manually administered drawing archive. Often, a major motive for becoming involved in CAD is to improve a firm's image and since drawings are shown to customers, it is usual to spend a lot of money on a good plotter to produce them. Considerable effort and time is spent on ensuring that drawings are of as high a quality as possible by experimenting with various combinations of pens, papers and plotting speeds.

Plotters, like displays, can be of calligraphic or raster types. Broadly, calligraphic plotters produce an image by using pens while raster plotters produce one by electrostatic means. The main characteristics of both of these types are described.

Pen Plotters

Plotters vary in price from £200–£300 to £200 000–£300 000. The cheapest are of A3 or A4 size, while the more expensive are 2 × 3 metres or more. The high cost is not necessarily for increased complexity but for factors such as higher speed, accuracy, repeatability and reliability.

There are various types of pen plotter on the market, the most widely used being flat-bed and drum plotters. In a flat-bed plotter (Figure 2.9), the paper is pre-cut to size and held on a table which is usually horizontal, although there is at least one type which is wall-mounted vertically. The paper is held on the table electrostatically, by vacuum or by magnetic strips, or is just taped down. A sliding beam moves along the table and this beam carries a pen-carriage which can move independently to and fro along the length of the beam. The combination of the two movements means that the pen can draw lines at any angle although, just as on calligraphic screens, positions cannot be infinitesimally varied but have a fixed small resolution.

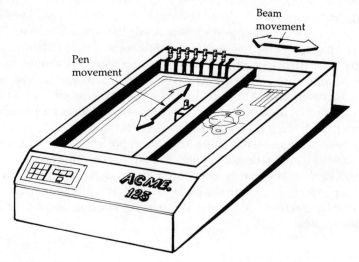

Figure 2.9   *Flat-bed plotter*

Very often, more than one pen is carried in a magazine - 4 or 8 is common. These can be of different colours or of different thicknesses. On CAD systems, pens are selected by software commands and it is useful to improve the appearance and readability of drawings by using different line widths for component outlines, dimension lines, hatching and other features of a drawing. The clarity of schematics, such as pipework layouts, can be enhanced by using pens of different colours. Some plotters are used for cutting out templates; they are then fitted not with a pen, but with a rotatable knifeblade. Others have a light source which can be used to expose photographic material. Since these, and others even more elaborate, are for specialized use, they will not be discussed further.

The cheaper plotters can only perform six basic operations: raise and lower the pen, and move one small unit in north, south, east or west direction. More expensive plotters can move in many more directions and in smaller increments. Even the cheaper plotters nowadays have micro-processors incorporated so that they can draw, in response to short commands, such features as lines, arcs, circles and alphanumeric text of various sizes and at specified angles.

The cheapest and the most expensive plotters are of the flat-bed type. Medium-cost plotters are often drum plotters (Figure 2.10) in which the paper is not pre-cut but is fed from a continuous roll. The feed roller and the take-up roller can rotate in both directions and so the paper can oscillate. Pens are held in a carriage which moves parallel with the axes of the rollers along a fixed beam, again enabling lines to be drawn in any direction. An advantage of the drum plotter over the flat-bed is that drawings, although of a fixed maximum width, can be of any length up to the length of the roll of paper. A disadvantage is that the same width is used regardless of the drawing size, which results in wasted stationery. In some systems, this is avoided by holding drawings in a plot queue and plotting them side by side, if they will fit.

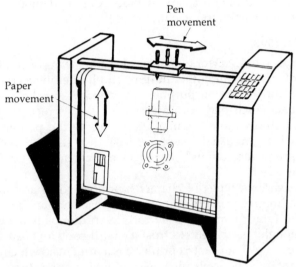

Figure 2.10   *Drum plotter*

There are some plotters which operate on the drum principle but use pre-cut paper which is held and moved to and fro by pinch rollers. Similarly, there are small flat-bed plotters which draw paper from a roll.

Electrostatic Plotters

Electrostatic plotters are becoming more and more popular because of their speed. The time taken to plot a drawing on a pen plotter depends on the amount of detail; an electrostatic plotter takes the same time regardless of the detail. A further advantage is that areas of the drawing can be blocked in; this is very useful in applications such as artwork preparation for printed circuit boards. Half-toning is also possible, areas being filled in with varying dot patterns, taking advantage of the spatial integration that we have mentioned before. Shaded pictures, like those found in newspapers, can be produced in this way.

On an electrostatic plotter the paper moves under an array of nibs, selected ones charging it negatively. It then moves through a bath of positively charged toner which sticks to the part which is negatively charged. On the best electrostatic plotters, there can be as many as 400 nibs per inch, overlapped so that no gaps occur in lines and extremely high quality plots result. The main disadvantage with electrostatic plotters is that a line drawing must be pre-processed in order to convert it to a raster image. This is called 'scan conversion'; this will be discussed in a later section.

The xerographic technique is also used for CAD hard copy, using devices much like the familiar office Xerox machine, but using a laser to charge the drum. It is possible, with multiple passes through different coloured toners, to obtain coloured plots. Coloured plots can also be obtained by ink-jet and other technologies.

Character Printers

Printers are used in CAD for two purposes: as a means of getting normal hard-copy of answers and other output, and for getting a crude copy of the drawing for discussion purposes by printing out the pixels which make up the screen display. The latter use is called 'screen dumping' and is used as a cheap though crude alternative to the plotted drawing. Plotted drawings are usually kept 'for best' - for sending to customers, storing as prime records and whenever a high quality but expensive drawing is needed.

There are various types of character printer: some are daisy-wheel printers, dot-matrix printers (these are 'impact printers'), and ink-jet printers. The type commonly used for screen dumps is the dot-matrix printer, although specialized electrostatic hard-copy units are also used.

A common type of dot-matrix printer has a print head which contains a vertical row of pins, typically nine, each of which can be fired separately to make a mark through a carbon ribbon on to a piece of paper. Printing is often done bidirectionally, while the carriage is moving from left to right and from right to left. In order to do this, the plotter has an internal print buffer in which at least a line of print can be held.

Speeds are upward of 120 characters/second in normal printing. The dots are placed on a 6×9 rectangle to form each character. Printing may be done on separate sheets of stationery using a 'cut-sheet feeder' or on continuous roll stationery using either friction feed, where the paper is trapped between rollers, or tractor feed, where the paper has holes which fit on to sprockets at its edges. The cut-sheet feeder and tractor feed are sometimes provided as optional extras.

The dot-matrix printer is very versatile, much more than, say, the daisy-wheel printer where the daisy-wheel must be changed to print in a different font. The dot-matrix type supports a variety of fonts and print styles: 'draft', which is fast but of minimal quality; 'near-letter quality' (or NLQ), which is slower but looks better, being formed by overprinting; Pica, a standard font; Elite, more compact than Pica; condensed; double-width; italic. Others are also available. Subscripts and superscripts are also possible. Different types of print style are selected by sending control codes to the printer. More expensive models have more pins in the head, sometimes as many as 24, and so have a higher print quality; they are also faster.

When a printer is being used as a hard-copy unit, none of these fonts is used. The printer is used in graphics mode and each pin fires to correspond with each pixel on the screen. The computer plays no part in the process, the dumping being just a matter between the display and the printer. An example of a screen dump is shown in Figure 2.11. It is uneconomic and unnecessary, in a multi-user CAD installation, to supply each workstation with its own screen dump device. Frequently, the displays are grouped into, say, fours and each group is served by one screen dump device which is connected to individual screens through a sort of electronic switch called a 'multiplexer'.

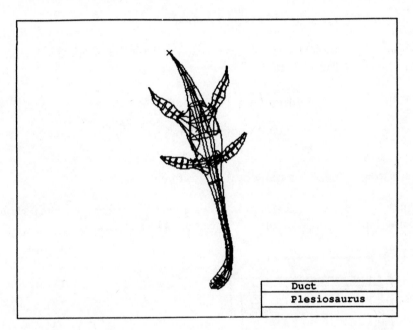

Duct
Plesiosaurus

Figure 2.11   *Screen dump*

Input devices

General Principles

The most direct way to communicate with a system is by spoken English or other natural language. Although this is not yet really possible unless the range of user utterances is tightly restricted, there is much research being carried out into the provision of efficient user–machine interfaces in general. There are many ways that a user can input commands and data to a CAD system (and we shall describe some of these in this section), but there is still no ideal input device for all users in all circumstances. For instance, it could not be claimed that a conventional computer keyboard is either natural or efficient. The main problem is the diversity of information that must be input to CAD systems.

Input devices can be considered in two ways: logically (by what the device does) and physically (by how it does it). The widely used Graphical Kernel System (GKS) classifies logical input devices as follows:

Choice   Allowing the user to select from a menu with a light pen, track-ball or similar device

Locator   Locating objects on a display

Pick   Indicating to the system when an object on the display has been located

String   Allowing alphanumeric sequences

Stroke   Allowing continuous drawing with a tablet, light pen or mouse

Valuator   Enabling numerical values to be input

All these are used in CAD applications and their variety accounts for the many types of physical input devices found in systems.

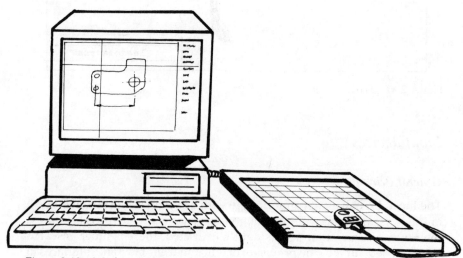

Figure 2.12 *Workstation*

Even on a simple system (Figure 2.12), several logical devices are needed.

The *menu* allows users to choose from a range of alternatives. Menu options are located by moving a *cross-hair cursor* and are picked by hitting a special *button*. It is possible to move the cursor by using the four arrow keys on the *keyboard* and picking by hitting the *space key*; this is inconvenient, so users are provided also with a *graphics tablet* and *puck*. The cursor mimics the movement of the puck and objects can be located by moving the cursor by puck movement. When the cursor has arrived at the desired position, picking is done by pressing a *button* on the puck. The *graphics tablet* can also act as a stroke device, for example, in inputting shapes from drawings taped to its surface. Dimensions can be input by using the *keyboard*.

We shall consider a sample of the many input devices currently on the market.

The Keyboard

The conventional or QWERTY keyboard is probably too familiar to need a great deal of explanation. Not only can the keyboard send alphanumeric and other characters that can be displayed on the screen, it can also send the non-displayable control characters which are part of the ASCII character set. Also, although these have standard names, for example, 'Start of header' (SOH - ASCII 1), 'Start of text' (STX - ASCII 2), which are meaningful in data communications work, their interpretation in CAD work is often set by the system being used. They are sent by pressing two keys at the same time: BEL - ASCII 7 rings a bell and is sent by pressing the control key and 'G' simultaneously. Some devices need a lot more control codes than the 32 provided by ASCII (dot-matrix printers are an example). The meaning of a character can be changed for this purpose by prefixing it with the control character 'Escape' (ESC - ASCII 27).

Users often have difficulty in going from one hardware system to another because many proprietary keyboards have their own layout: although the alphanumeric characters are always in the same place, other characters such as '+' are in different positions. There have been attempts, not very successful, to improve the ergonomic layout of the keyboard. One such has the keys arranged in an arc suiting the natural sweep of the hands and the keys ordered in a more convenient way than the familiar QWERTY. Another attempted improvement is the Chord keyboard which has just five keys and so can be used with the fingers of one hand.

Most modern keyboards are supplied with a set of programmable function keys (PFKs) which the user can program so that, when they are pressed, a short command (such as 'Help') is sent to the system.

The Graphics Tablet

Graphics tablets (Figure 2.13) are small (280 × 280 mm is typical) flat plates which may be used for various purposes. Larger, more accurate versions are called 'Digitizers' and are used for entering drawings such as maps and schematics directly into a system. Graphics tablets (or 'bit pads') can work in several ways. A popular system is to have a grid of wires sandwiched between two plates. A coil is carried by a puck or stylus which can be moved over the surface of the tablet. This generates voltages in the wires of the grid and, by analysis of these, the position of the puck or stylus can be found relative to some reference point. Some tablets are provided with a digital read-out showing the coordinates.

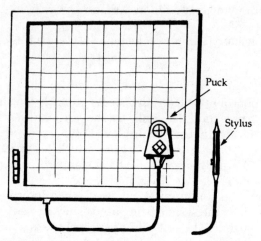

Figure 2.13   *Graphics tablet*

Graphics tablets may be used in various ways:

to drive the screen cursor which mimics the movement of the puck or cursor;
to support a menu card from which options may be picked;
to input shapes from drawings taped to the flat surface: points on the drawing may be picked or can be taken in automatically in a stream, the sampling being done at intervals of time or of distance;
to provide a natural means of sketching for users of painting systems.

There are available similar devices for input in three dimensions rather than from a flat surface. These are not yet widely used in CAD but they might be useful for input of three-dimensional coordinates from physical models to computer modelling systems.

The Light Pen

Light pens (Figure 2.14) are used for referring to points and objects on the graphics display directly. They were once very popular but are less commonly met nowadays. They can be used in various ways: for choosing from screen-displayed menu options, for picking objects on the screen and moving them about as though they are adhering to the pen tip and for sketching directly on to the screen. They have the advantage that users can keep their eyes on the screen at all times. There exist touch-sensitive screens that are used in much the same way except that the tip of the finger is used instead of a pen.

Figure 2.14 *Light pen*

Drawing Scanning Devices

When a firm installs a computer-aided draughting system, it must make a decision about the old manually-produced drawings. Should they be left alone and two systems of drawing office records maintained - a messy solution? Or should all the old drawings be entered into the new system by redrawing - often an immense task? An ideal solution would be to enter the drawings into the system automatically, resulting in a unified system with little waste of time. There are available devices that can do this. They work by scanning the paper drawing and holding it in raster form, then converting it back to calligraphic form for input to the system. They are currently very expensive.

There is another difficulty also. A computer-aided draughted drawing usually has more information than the mere lines that appear on the visual representation. For instance, most firms that are seriously involved in CAD have standards for symbols, layer utilization and so on. It is impossible to input drawings automatically in a satisfactory way without a good deal of human post-processing of the scanned version and this reduces the benefit of the system considerably. Even so, these devices are at an early stage of development and when the difficulty mentioned has been resolved, they should become invaluable.

Other Devices

A further selection (Figures 2.15–2.18) of input devices found in CAD systems is listed below. They will not be discussed in any detail.

Mouse   Potentiometers are coupled to a roller ball so that
                     when the mouse moves on a surface, its movement
                     can be transmitted to the system. They are used like
                     the puck on the graphics tablet.

Track-ball   These work like the mouse, but the action is inverted,
                     the ball being manually rotated.

Joystick   Superior versions of the ones used on computer
                     arcade games and used in much the same way.

Thumbwheels   Two knurled wheels mounted on the keyboard which
                     are operated by the thumb and finger of one hand.
                     One wheel governs horizontal movement of the
                     cursor and the other vertical movement.

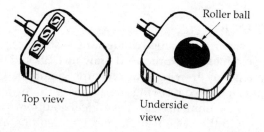

Figure 2.15 *Mouse*

Figure 2.16 *Trackball*

Figure 2.17 *Joystick*

Figure 2.18 *Thumbwheels*

Storage

General Principles

In any computer system, we need to store data and programs. Computers have several types of storage, each having cost, size and speed appropriate for different purposes. Examples are:

Internal storage
   Registers
   Random access memory
   Read only memory

External storage
   Drum store
   Disk store
   Tape store

We shall describe the types of store with which CAD users are concerned.

Internal Storage

This is the storage needed for holding data and programs temporarily while the system is being used. It is organized as a set of fixed-length words, lengths being 8, 16 or 32 bits in the case of microcomputers but any length in other kinds of computer. It is sometimes called 'random access memory' or RAM; stored items are referred to by their addresses and accessed in a fixed time. RAM can be written to as well as read. It is now mainly semi-conductor store – the contents of this type of store disappear when the power is switched off: it is said to be 'volatile'. Because semi-conductor store is volatile, its contents must be maintained either by a constantly applied current ('static' RAM) or by periodic refreshing ('dynamic' RAM).

For some purposes, it is necessary to store information permanently. This is done by using 'read only memory' or ROM, which can be read but not overwritten. ROMs are used in many applications, an example is the storage of the different print styles of a dot-matrix printer.

In the simplest computers, such as early microcomputers, the words of memory are organized sequentially, one word after another. The programs must either be shorter than the total memory size available or be split up by the programmer into chunks called 'overlays' which are loaded to main store from backing store as required, the program being run piecemeal. This simple system is quite unsuitable for a multi-user system because of the need to load each user's program as their turn to use the computer comes round. Even if they were all using the same program such as a draughting system it would still not be satisfactory because, although the one program could be organized so that it was shareable, each user would need their own workspace to hold partially completed drawings.

This difficulty is resolved by using 'paging'. Main storage is organized as blocks or 'page frames' of a fixed size into which can be loaded parts of programs or bulk data which is divided into blocks of the same size, called 'pages'. The pages of a program do not need to be stored consecutively but can be loaded into any free page frame. The bookkeeping is done by hardware and the user need know nothing of the way the system is organized. Thus, several programs, or parts of programs, or users' workspaces can exist in the computer at the same time. A similar scheme is used in 'segmented' storage - the difference being that the program is not split into equal sized pages but into segments which may be of different sizes. The division into pages is a physical operation, there is no logical significance about the way the program is split up and so the system can be transparent to the user. In contrast, the division into segments is done on a logical basis by the user. Some computer systems, and the PRIME is one, use a combination of both paging and segmentation. These have quite complex architectures.

In multi-user systems there is often no theoretical restriction on the number of users that can be served. There are practical restrictions, for instance, the maximum number of lines that can enter the system; also, there is an upper limit to the number of users because if too many are allowed on the system, response time deteriorates.

External Storage

As well as main storage, used for temporary storage of programs and data, all systems have backing store which is used for longer-term storage. Many types of backing store are used, and these are mainly magnetic. We shall describe the ones most widely met in CAD systems.

Magnetic Disk (Figure 2.19)

Usually, large computers use magnetic disks for bulk storage. These disks are of two types: exchangeable, in which disk-packs may be taken out of the drive and replaced, and fixed disks, in which the unit is sealed. Information is stored on the surfaces of a number of platters which are mounted on a central spindle, much like an old-fashioned jukebox. A comb of read/write heads can move radially in and out of the spaces between the platters. Information is held physically on the disk in concentric tracks which are divided into a number of sectors. A block of information can then be accessed by quoting platter surface number, track and sector number. Logically, the data is organized into cylinders, which results in faster access because the read/write heads cannot move relative to one another. The read/write heads do not touch the rigid disks but float on a very thin air-cushion close to the disk surfaces. Magnetic disks can hold, typically, 300 Mbytes and because of their high rotational speed can access data very quickly.

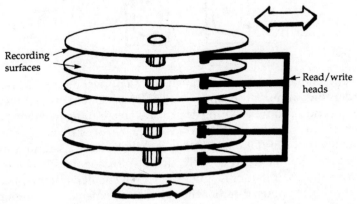

Figure 2.19   *Magnetic disk*

Magnetic Tape (Figure 2.20)

Magnetic disks are direct-access devices - the read/write heads can move to any track directly, which contributes greatly to their high speed of access. However, it is often not necessary to access data in this way and it is sufficient to access it 'serially'. If we wish to play a track from a long-playing record then, assuming that we can locate it from the sleeve and empty tracks on the platter, we can move the stylus directly to the correct position. If we wish to do the same with a tape, then we must move through the tape from the beginning to the desired point. The first is direct access, the second serial access. Note that if we wanted to play all the tracks, then it would not matter which of the two was used. Magnetic tape is used for storing data when it does not matter that it can only be accessed serially. An example is when a security copy (or 'back-up') of a magnetic disk is made.

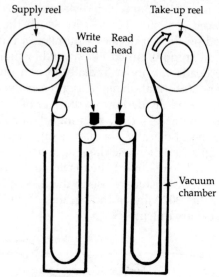

Figure 2.20   *Magnetic tape*

Magnetic tape drives work in much the same way as reel-to-reel tape recorders but, because the tape moves much faster and often does a lot of stopping and starting, a 'spring' is provided by draping two loops of tape into vacuum chambers; this avoids damage or stretch of the tape. Magnetic tapes can hold as much as 50 Mbytes and, assuming that the information can be read serially, have a reasonable transfer rate - approximately half that of magnetic disk. On smaller systems, magnetic tape cartridges are used rather than the more expensive magnetic tape; these are called 'tape streamers'. They are used for the same jobs as tape - for holding security copies and other serially organized data and as a means of data transfer between separate installations.

Floppy Disks

Various sizes of disk are used on microcomputer systems: $3^1/_2$ and $5^1/_4$ inches (89 and 134 mm) are now the most commonly found. The latter is the familiar 'floppy disk', the former is the 'micro-floppy' which is becoming more and more popular.

The floppy disk (Figure 2.21) consists of a single plastic disk, coated with a magnetizable material, which is rotatable within a square card or plastic protective cover. The recording surface (or surfaces, if the disk is double-sided) can be accessed by the drive's read/write head through a slot in the cover. The data on a floppy disk is organized into tracks and sectors and there is a hole through the disk which serves to identify the first addressed sector to the disk controller. This type of disk is called 'soft-sectored'; a 'hard-sectored' version was once popular: this had timing holes at the beginning of each sector. Microcomputers usually have one or two floppy disk drives into the slots of which the disk is 'posted'. It is trapped there and rotated by a clutch arrangement. The read/write head straddles the disk surface and moves radially to a desired position, being driven by a stepper motor.

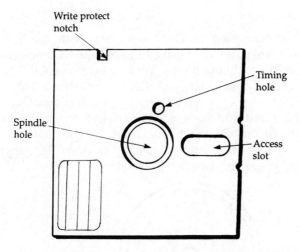

Figure 2.21   *Floppy disk*

The capacity of floppies varies; a typical one has two sides, 40 tracks per side, 9 sectors per track and 512 bytes per sector, giving a capacity of 360 kbytes.

Micro-floppies (Figure 2.22) are more robust and reliable than the ordinary $5^1/_4$ inch floppy. They are physically much less flexible and the recording surface access slot is protected by a metal sliding shutter. They also can hold a fair amount of data, a double-sided one holding typically 720 kbytes. This capacity can be compared with the size of a typical novel which is about 500 000 characters, or 500 kbytes. A disk of this capacity would probably be capable of holding twenty or thirty reasonably detailed drawings.

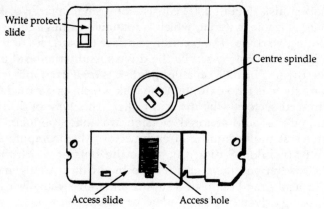

Figure 2.22   *Diskette*

Hard Disks (Figure 2.23)

Most microcomputers used in CAD are fitted with hard, fixed or Winchester disks. These are not visible from outside the microcomputer but are sealed inside it, although some are now exchangeable. The disks, which are rigid, have a much higher rotational speed than a floppy and so transfer data a lot faster. They have one or two platters which are single-sided or double-sided and have a comparatively high capacity. On cheaper micros, 10 and 20 Mbytes is usual; on the more expensive workstations, Winchesters with a capacity of over 100 Mbytes are available and capacities are rising very rapidly.

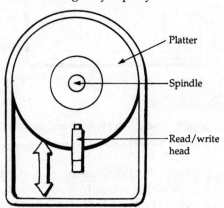

Figure 2.23   *Winchester disk*

Processors

Mainframe, Mini and Micro

CAD programs need powerful processors. On a well-designed system, using graphically displayed output, it is easy to imagine that there is little calculation being performed. Good systems are like swans swimming on a pond: on the surface they are moving leisurely and gracefully; under the water, their legs are paddling away furiously. Even operations which look straightforward - deleting a line on a computer-generated drawing is an example - involve a considerable amount of processing. Since users expect the system to respond to commands rapidly (research indicates that any wait over 2 seconds is regarded as intolerable by most users), it is important that the processor is capable of providing this degree of operational speed. It is also essential that the computer should be able to hold numbers to a high degree of accuracy.

A few years ago, these requirements meant that CAD could only have been done successfully on the larger and more expensive computers; to use microcomputers was unthinkable. However, today 16 and 32-bit microcomputers are commonly used for CAD. The distinction between mainframes, minicomputers and microcomputers is now blurred and it may soon be meaningless to use this kind of classification. What was meant by the three types has never been very clear from a physical point of view; there were attempted definitions such as:

'Micros are of 8-bit word length'

'Minis are of 16-bit word length'

'Mainframes are of 32+ bit word length'

but these do not now apply. It is perhaps possible to write down a loose definition by function.

Mainframes are expensive computers used when there are many users and the unit cost is lower if they share a computer between them; or when it is necessary for users to have access to a wide range of different systems; or when they need to have access to a large centralized database. They are also used when fast complex calculation is wanted - for instance, when large three-dimensional finite element analysis models are to be processed.

Microcomputers are the cheapest kind of computer. They range from the simplest type of processor found on the personal computer up to the modern 32-bit machine. They are becoming increasingly popular in the engineering workstation, which has already been briefly discussed. Very often, when micros are used for CAD, their performance is enhanced by fitting a second processor (or co-processor) to handle the arithmetic on floating point numbers. And, as we have mentioned, graphics capability can be enhanced by adding one of a number of proprietary graphics boards.

Minis lie between the two extremes of desktop microcomputer and large powerful mainframe. The larger ones are called 'superminis'. They are widely used for CAD work.

Organization of CAD Processors

The simple type of organization is when each user has a workstation and the workstations are not connected together in any way. These are then called 'stand-alone' systems and are found typically in small firms where they are often used for draughting. They are merely replacements for drawing boards and, as in most cases when a computer is used purely as a replacement for a manual procedure, no value is added and the computer does not live up to the firm's expectations.

An obvious problem is the storage and retrieval of drawings. Typically, on a stand-alone system, drawings are stored on a hard disk; since even a modest drawing takes up a surprising amount of storage, the largest disk becomes filled quickly. Even if it were possible to use the hard disk for drawing storage, it would restrict users to using the same physical system at all stages of a drawing's life. In practice, drawings are held on floppy disks or cartridges to free users from this constraint, also (and more importantly) to keep security copies of drawings. The result is that a drawing register must be kept, just as is done with the manual system. This must be rigorously maintained since, apart from the sticky label, it is impossible to tell what is on a disk by just looking at it. Often, the prime drawn component record is a plotted drawing. Although there are computer systems to do this sort of Drawing Office administration, keeping track of the location of drawings on disks and in filing cabinets, very little has been gained over the manual system.

It is, then, almost essential that workstations should be connected so that information can be shared between users. Historically, the first way of doing this was by using a centralized multi-user system in which users were connected to a centrally located, usually large, computer. The advantages are many. Drawings can be archived centrally on large bulk storage with security copying being done in a systematic way; the issue of drawings to users can be controlled so that users can be confident that they are working to the correct issue; users can have access to centralized standards and information such as material specifications; strict drawing standards can be imposed by using standard configurations for elements such as dimension lines. And, very importantly, drawings can be passed to other departments, such as Production Planning, in an organized way.

There are, however, some drawbacks to this type of system. One is that the extra complication often means that systems are hard to use for the beginner and a longer initial training period is needed to attain satisfactory performance. Multi-user systems run with an operating system overhead and it is not uncommon to find a machine which spends some of the best microseconds of its life running its operating system. There is also a staff overhead, since computing staff are needed to organize the machine, to set up running procedures and to operate it.

There are technical problems, too. The most obvious one is that if the central computer fails then the system cannot be used by anyone. It is necessary to have an expensive maintenance contract so that system down-time is minimized. Another snag is that response time is not linearly related to the number of users. If we gradually increase the number of users on a system, response will be satisfactory up to a point, but will then deteriorate very rapidly. The point of rapid deterioration will depend on the usage profile and so it is difficult to plan for expansion.

Use of Networks in CAD

As has been discussed in the preceding section, using isolated systems is neither convenient nor effective. Users must be able to communicate to get the best out of a system.

Intercommunication, or 'networking', can be done in various ways, the methods of connection being called 'network topologies'. Physically, where the stations (or nodes) of a network are geographically close, the connections can be twisted pairs of wires or co-axial cable; the network is a local area network, or LAN. For networks that are more widely dispersed, nodes can be connected by, say, telephone lines. This is a wide area network or WAN.

There are several popular network topologies, each with its own advantages.

The Star or Hierarchic Network

The centralized system described in the preceding section is a simple example of the star topology, although the term 'network' is usually reserved for cases where nodes are provided with processors - for instance engineering workstations. This is a logical extension of the modern move towards distributed intelligence which has been mentioned before. A property of the star network is that there is only one path between each node, so, if for any reason part of the path breaks down, there can be no communication between the end nodes. The most robust topology is to have each node interconnected to all the others (Figure 2.24) but this is an expensive solution, and one difficult to control.

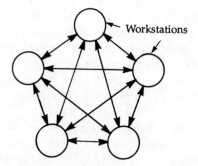

Figure 2.24   *Fully connected star network*

The Ring Network (Figure 2.25)

This goes some of the way towards the fully-connected topology. In this configuration, messages are circulated around the nodes until the correct destination node is reached. The ring network is basically democratic, and each node can have the same importance as any other; not only the processes but also the information can be distributed around the ring. Often, this symmetry is not maintained and one or more nodes are selected as main stores of information. These may not just be work-stations, but larger processors, for example, superminis. It is also economic to locate expensive peripherals such as plotters at one node which can be delegated to deal with the plotting of drawings. A disadvantage of this topology is the long mean communication path between nodes.

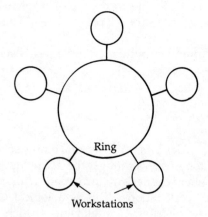

Figure 2.25   *Ring network*

The Bus Network (Figure 2.26)

This topology uses a single cable for the transmission of messages. Nodes are hung on this cable, or 'bus'. A message is placed on the bus by the transmitting node and is carried to all the nodes on the line using a carrier signal. The system is complicated by the need to resolve collisions, when two nodes try to broadcast a message simultaneously.

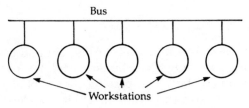

Figure 2.26 *Bus network*

A practical network may include elements of all these topologies, for instance, ring networks may be connected by 'gates' and may be joined to a bus network.

Engineering Workstations

Over the last few years there has been a continuous move towards 'distributed intelligence' in computer systems. Originally all the processing power of a system lay in the central computer, the peripherals being merely slaves with little or no capacity for doing anything beyond what they were instructed to do by the computer. They were known as 'unintelligent' or 'dumb' devices. Today's graphics displays are provided with integral display processors which are capable of doing some of the more specialized tasks needed in graphics. Special-purpose chips known as 'Geometry engines' have been developed to assist in the process. Graphics operations are discussed in a later section. A selection of the commoner operations that today's 'smarter' graphics displays can perform are:

Clipping pictures to suit the display area
Block-filling polygons with predefined patterns
Anti-aliasing, which has already been mentioned
Storing whole drawings, so that redrawing is fast
Storing segments of drawings, which can then be copied and moved
Assisting the processor in hidden surface removal

Even personal computers with their dumbest of monitors can be fitted with graphics boards which have their own frame buffer to improve screen resolution and provide some graphics processing facilities.

An extreme example of distributed intelligence is the engineering workstation. This is usually a powerful microcomputer, often 32-bit, which has a large internal memory, has high resolution graphics and has its own auxiliary storage, usually a high-capacity Winchester and possibly a floppy disk drive. The workstation can be used as a stand-alone computer but at its most useful is connected to other workstations and processors in a network.

Problems

1. An average typist can perform at 40 words per minute. Assuming that the average word (including the delimiting space) in normal English is about six letters long (in fact, it is about 6.1 for this book - check it if you wish!), then the average word will take up 60 bits - (six letters) times (eight data bits + one start bit + one stop bit) - if expressed in ASCII code. This means that the average typist transmits at about 40 baud. Normally, the minimum rate of system transmission is 1200 baud.

Explain this discrepancy.

2. The following terms are used in data transmission. Write a short definition of each in your personal CAD dictionary.

(a) Stop bit
(b) Start bit
(c) Protocols
(d) Longitudinal parity
(e) Check numbers
(f) Data encryption
(g) Hamming codes
(h) Modems
(i) Packet switching
(j) DATEL

3. In Morse code, letters are represented by codes of different lengths. The commonest letter found in normal text, 'E', is represented by '.'; the rarest letter, 'J', is represented by '·----'. In many applications this leads to a considerable saving in transmission time because of the compression of the commoner letters. In ASCII code, all letters have the same number of bits.

Investigate the possibility of using variable length letter codes in computer transmission. (Note that a considerable amount of work has been done on this.)

4. Professor Ben Trovato, that well-known inventor, suggests to you that a useful application for computer graphics is in the car rally field. A portable computer could be mounted on the dashboard of the car, and a highly detailed map of Great Britain held on the hard disk. Starting at any grid square, the screen display could be zoomed and panned, so avoiding

the problem of map folds. Furthermore, the computerized map could be traced with a light pen to give average speeds. Because of the wide range of colours that can be displayed on modern screens, a detailed relief map could be provided. It might also be possible to display a three-dimensional model... etc.

Investigate the feasibility of the scheme and write a short note on your findings to the good Professor. You might find it useful to start by estimating the resolution needed to display a grid square of the same quality as an Ordnance Survey map, and then to calculate the amount of storage needed to hold a representation of the square.

5. Why does a colour television set display pictures of far higher quality than does a computer screen, when computer colour displays often have as many as five times more pixels than do colour television sets?

6. Professor Ben Trovato has suggested that the most convenient method of locating items on a screen is to pick up the eye movement of the operator and transmit it to the computer.

Discuss if this is feasible and, if so, how it might be done.

3 Software

Introduction

'Caught in the Draught'

Scenario

Coolpoint make domestic washing machines. They are rather a traditional firm and have not been doing well lately in a highly competitive market. The Managing Director has recently learned that their major rivals, Watussi, have installed a large CADCAM system and has called an informal meeting to discuss whether they should follow suit. Present are the Engineering Director, the Marketing Executive, the Chief Designer and the Computer Manager. They have removed their jackets and draped them over the backs of their chairs to show evidence of dynamism and willingness to get on with the matter in hand:

MD: As you know, we have been taking a beating from Watussi since they developed their Dual Non-tangle Laundry Centre. I can best show you the situation by this graph of sales over the last six months. [*Displays chart with red line sloping down at an approximate angle of thirty degrees.*] We know that our marketing is at least as good as theirs. But we have been beaten to the punch consistently over the last two years. Need I mention the Watussi Wispa-Quiet? [*Sharp intake of breath all round the table.*] We had as advanced a model under development at the same time but their version hit the high street three months before ours. It seems to me that it is vital for the viability of the firm that the lead-time between concept and production be drastically reduced. I have got you all together to sound you out about our possible future involvement in CAD.

ME: They have certainly up-marketed their corporate image since they got their system. Have you seen their latest TV ad? You know, the one with Wispa-Quiets arranged in a circle like Stonehenge around a computer and a voice-over saying 'From the dawn of time, Man has been a calculating animal...'.

ED: I suppose our real aim should be to increase Drawing Office productivity. I read somewhere recently that you can expect to increase drawing throughput by a factor of three to ten by using CAD. We could cut costs by shedding some of our draughtsmen.

MD: True. But the cost of running the DO only counts for about two per cent of our total costs. Even if we did achieve the figure you mentioned, it wouldn't have much effect on the cost of our product. Perhaps it might be better if we left that one for a while. Can we concentrate on cutting down product development time?

CD: If we could speed up detail draughting, we could get to the prototype stage much faster and this would give us more time to develop the product fully. Then we wouldn't hit so many snags in production.

CM: I have heard that CAD is only useful where a lot of modifications have to be carried out on drawings.

CD: It should be invaluable for us, then. Our new models are normally only slightly different from the previous ones. We are only talking about draughting, though, and CAD involves a lot more than that. I have heard that Watussi have spent close to a million pounds on their system, which is integrated with production.

CM: They are also using finite element analysis, anthropometric modelling, a specialized sheet metal design system…

CD: I don't think that we should rush into this headlong. Perhaps you might remember the disruption when we went metric. I can't see our designers, good as they are, taking to such a radical change easily.

MD: Well, I doubt if we could afford anything like the money that they have spent, in any case. We could not run to anything like that amount. Let's just start by considering draughting alone. We can always expand our involvement later on. Is everyone agreed that we should explore the idea further? [*Chorus of assent*.] Can you all go away and think about it? In the meantime, I shall invite some vendors to come in and show us their wares.

■   ■   ■

In this section and those which follow, we shall be describing some of the major subsystems which make up a full CAD system. Some of these have been mentioned by Coolpoint's Computer Manager but no one else seemed to appreciate that CAD was concerned with anything other than draughting. In fact, many experts consider that traditional engineering draughting might well disappear in the future, being superseded by three-dimensional modelling. In order to understand the true place of CAD in design, it is useful to spend a little time considering the design process.

The Design Process

CAD is the use of a set of programs which assist in some stages of the design process. In order to provide programs which will be useful to the designer, it would be profitable to know what design is, what are the functions involved and how the designer works. Unfortunately, the role of computers in the design process is still the subject of research. As in many other computer applications, it has not been considered useful to wait until a theoretical base has been sorted out; a system that is 90 per cent perfect is thought to be better than no system at all. We shall take a simple view here and define the design process by listing what many observers think that designers do . It is true to say, however, that many successful designers do not work in such a systematic way.

1. The designer is provided with a 'brief'. This can come from various sources - for instance, from the Sales Department as a result of a market survey or from customers' complaints about an existing product. The brief is a broad statement that there is a demand for a new product to perform a particular function. For example, the brief might be a request for the design of a novel toothbrush.
2. The designer then prepares a 'design specification' which is a detailed list of the factors to be taken into account when the design is created. Some of these factors may be essential: others may be merely desirable. The design specification may include items such as cost range, safety standards and details of the target buyer. Many factors are conflicting and involve compromise - for example, in designing an aircraft structural member, it is essential that it should be strong enough and yet it is also a pressing requirement that the weight should be as low as possible. This sort of compromise always occurs in design and a skilled designer will be accustomed to resolving it satisfactorily.
3. The designer usually conceives several solutions to the design problem in sufficient detail for them to be compared. This is the main inventive stage and designers perform it in many different ways.

Techniques have been devised to assist in the generation of ideas (brainstorming is an example) but, essentially, the success of this stage depends on the creativity of the designer.

4. The possible solutions are considered and comparisons are made between them. Usually, one or two of the designs will be found to show more potential for satisfying the specification. These are selected for in-depth analysis.

5. The 'best' designs are analysed in detail in order to make sure that they do, indeed, match the necessary requirements. The original ideas are normally quite sketchy; it is necessary to translate these to a realistic product. So the performance must be analysed, the load-bearing parts must be stressed, materials must be selected, decorative finishes must be considered and so on. Much of the work done at this stage is calculation and involves little creativity.

6. The design is detailed for prototype manufacture. Detail design is a long and skilled process and the success of a product often depends on its detail design. The prototype is tested against the specification and, if satisfactory, the design proceeds to manufacture.

At all stages in the process, decisions are made about the correctness of the design. If necessary, the designer will return to a previous stage in order to rectify some shortcoming in the design. These trial-and-error loops are known as 'iterations' and it is usual to perform many of them during the design process. If a skilled designer were allowed unlimited time to develop a product, then something very close to a 'best' design would probably result but it would need many iterations. In practice, it is impossible to allow unlimited time, not just because of the labour cost involved but also because it is usually an advantage to put a product on the market as soon as possible.

Computer-aided Design

The processes involved in design can be divided into two broad types of activity: synthesis and analysis. Synthesis is the creative part of the process - the generation of ideas. Analysis is the evaluation of these ideas - finding out whether the design fulfils its objectives, is strong enough to take the loads imposed on it, fits inside the space available, performs efficiently enough, is sufficiently attractive to potential buyers, is convenient and safe for users, does not harm the environment and so on.

Currently, computer systems are useful aids in the more analytic parts of the design process. Some examples of their use are:

3-D modellers	Used to define the geometry of a part, to view a design from different angles, to calculate weight and other mass properties and to detect collision with neighbouring parts

Finite element analysis	Used to calculate stresses and deflections in a loaded component
Anthropometric modellers	Used to check whether the design matches human measurements
Materials selection systems	Used to pick materials with required properties
Machine-element design systems	Used for the design of components such as fixing devices, bearings, gears, cams etc
Draughting systems	Used for the preparation of conventional engineering drawings and parts lists

This list is confined to mechanical design; a similar one could be devised for electrical and electronic design.

None of these systems is designed to assist creative activities; they are all useful in the *development* of ideas rather than in the *generation* of ideas. It might be argued that 3-D modellers, when used to view designs, are equivalent to the industrial designer's sketchpad. However, current systems suffer from some drawbacks. A given system can only model some of the possible shapes in which a designer may be interested. The designer is already working under extremely difficult constraints - for instance, in trying to design an object of infinite worth for zero cost. The further constraint that only certain shapes are allowed by the system is a serious obstacle to the creation of a good design.

Another problem is that most systems now available do not permit the interactive design of shapes and so designers are forced to work in a way that does not come naturally to them. Usually, when designers first create their solutions to a problem they communicate them in quite a sketchy way, omitting much detail, neglecting dimensional exactness and so on. Most three-dimensional modellers require an exact definition of the form of the model so that the designer is forced to consider exact proportions at an early stage. For some time, it is likely that designers will still prefer to 'sketch on the backs of envelopes'. This, however, is probably only a temporary problem which will disappear with the emergence of new graphics systems and there are systems recently developed which claim to permit a more natural way of working.

Three-dimensional modelling is still at an early stage of development and a lot of money is being spent on research into the field; it is very likely that future 3-D modellers will be useful aids in the visual design of products. It is also likely that other aids will be developed to assist designers in the generation of novel ideas and that artificial intelligence will play a part in these aids.

Draughting Systems

General Principles

Since draughting systems are the most common of CAD subsystems, they will be treated in some detail. Though there are some experts who have for years predicted that two-dimensional drawing will disappear as an engineering activity, giving place to three-dimensional modelling, it does not seem to be happening to any degree. Their main arguments against conventional drawings are:

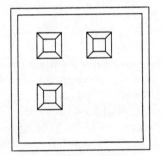 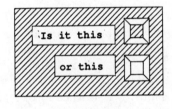

Figure 3.1   *Ambiguous body*

(a)   Two-dimensional drawings can be ambiguous. Figure 3.1 is commonly used as an example of a drawing which is perfectly legal, but does not define a unique geometric form. The section in the middle can be either a hole or solid.

(b)   There is no check on whether the form shown can exist at all.

(c)   Often, the conventional engineering drawing does not completely define the form of the component being drawn. In casting drawings especially, some of the definition is left up to the patternmaker. In three-dimensional modelling, all the form must be defined for the model to exist at all.

(d)   Three-dimensional modelling is superior because a conventional engineering drawing can be obtained by projection from a three-dimensional model but a three-dimensional model cannot be directly obtained from a conventional engineering drawing.

(e)   Mass properties such as volume and position of the centroid cannot be calculated easily from a two-dimensional drawing; they can be calculated automatically from a three-dimensional model.

We would suggest that some of these arguments are suspect and might be answered as follows:

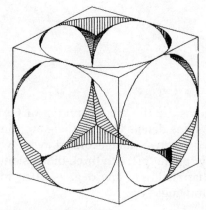

Figure 3.2   *Impracticable solid model*

(a)   Any competent draughtsperson would give extra information such as a section to remove any ambiguity from a conventional drawing. In a view where it is not clear whether a hole passes through the part, it is customary to add a note stating that the hole is 'through'.

(b)   Figure 3.2 shows a form that may exist on a screen, but not in real life. It has been produced on a solid modeller which has not had the sense to tell that it would fall into eight pieces if it were manufactured.

(c)   If dimensions are left off a drawing, it is usually because they are functionally irrelevant. In any case, most three-dimensional models do not fully define the form, such details as blend radii being omitted (in some cases, because they are impossible to model).

(d)   So what?

(e)   It is impossible to disagree with the last criticism of two-dimensional drawing; however, on some three-dimensional systems the values obtained for mass properties are so wildly inaccurate as to be useless.

It may well be that three-dimensional modellers will become the main way to define components' geometry in the future, but they will have to be considerably more advanced than those offered at the moment. This is not to say that they have an unimportant role in CAD and their uses will be discussed in a later section. None of this criticism applies to surface modelling which has very clear advantages over manual methods in suitable applications.

Three aspects of computer-aided draughting systems will be examined:

1 the user interface, because to a large extent it determines how easily a system is used;
2 elementary drawing operations, because they are the most frequently used operations by far and have a great influence on the speed with which a new user can get to grips with the system;
3 advanced facilities, because the increase in drawing productivity depends on them.

There is no standard set of operations that is common to all systems. All can be used to draw lines and arcs, write text, and for the other operations which directly replace manual procedures, but each of the many available systems has its own particular style and its unique set (often hundreds) of commands. Draughting systems, like all CAD software, are in a constant state of development with frequent new issues. With each release there are added features (and fixes of the bugs from the previous release). The added facilities are often suggested by user groups and are often aimed at specialized fields. The operator, faced with a bewildering battery of options, commonly settles on a working subset, although some users take it as a point of pride to know all the commands on their system.

Three draughting systems are used here for illustration:

AUTODESK's AutoCAD
PAFEC's DOGS
DELTACAM's DUCTdraft

This is not a comparative study of draughting systems, and these have been chosen because of their differences; AutoCAD is a widely available and popular general-purpose micro-based system. DOGS, although now available on micros, was originally designed to run on minis like VAX and PRIME computers. DUCTdraft is an example of an up-to-date workstation-based system using modern display techniques to good effect. No attempt will be made to make any judgement of the comparative *quality* of the three; like many of today's draughting systems, each is of a different style but well thought out and easy to use. Most users are loyal to their own system and hypercritical of others, but, unlike the situation a few years ago, many of today's systems are of a satisfactory quality. It seems that, at last, systems designers are becoming aware of users' needs, in draughting systems at least. The same does not apply to some of the other elements of CAD where user interfaces have often been designed inconsiderately.

The User Interface

The most obvious way that one draughting system differs from another is in its appearance to the user. Draughting systems are invariably menu-driven, the menus being either displayed on the screen, which cuts down the drawing display area or necessitates a separate monitor, or being printed on card and taped to the surface of a graphics tablet – forcing the user often to switch gaze from the working drawing, which is distracting. The commands can also be typed, usually in a concise form, and many of the faster users work with a combination of menu selection and keyboard input; it is a feature of a well-designed system that users can, to some degree, adjust the way of using the system to suit their own preferences. A cross-hair cursor is displayed on the screen and it can be positioned over an item and a button pressed to transmit the user's intention to the computer. The menu item under the cursor is highlighted by temporarily being displayed in reverse video, in a distinctive colour, or by being underlined.

Menu organization differs from system to system. DOGS places the whole range of menu options in a vertical strip down the left-hand side of the screen. Each option consists of two parts, a menu option, which is a mnemonic word, such as 'LINE', and a submenu item which is a number. For instance, 'Delete a line' is chosen by first picking the menu item 'DELETE', then the number 2. Because of the large number of operations (getting on for three hundred) available, a good deal of memorizing is necessary before the system can be used with any fluency. This does not seem to bother DOGS users.

AutoCAD uses a system of 'pull-down' menus. A small root-menu is displayed; this contains the main categories of commands, such as 'DRAW'. If an item on the root menu is picked, then a further submenu is displayed and so on. This is slower than the DOGS single-level menu, but does not need anywhere near the same feat of memory.

DUCTdraft combines the conciseness of DOGS with AutoCAD's ease of use. A fairly comprehensive on-screen menu is displayed and, for some items, small pop-up menus are provided.

Many systems now use pop-up menus; some of these follow modern taste and have menu items displayed as icons. These certainly help new users but tend to irritate the more skilled. When menu cards are used, the card is divided into rectangular areas; these are also usually iconized. Often, only a selection of the commands is printed on the menu card and many systems permit users to define their own menu layouts which they may specialize to their own working preferences and fields of application. Although many of the large-scale systems use menu cards, they seem to be giving way to on-screen menus for general use.

When commands are entered from the keyboard, it is sometimes possible to use abbreviated forms of the full command. On DOGS, if the letter 'K' is typed, then what follows is interpreted as a command. 'KDE2' is a short form of 'DELETE 2', which means 'Delete a line'. This is rather more convenient than it sounds. In AutoCAD, the treelike structure of submenus can be bypassed by typed commands.

All respectable systems provide intelligible prompts and help facilities. Prompts can often be set to various levels - in DOGS, 'novice' and 'expert' prompts are available and can be selected on entry to the system, expert prompts being terser than novice prompts. All systems provide some sort of help when operators do not know what to do next. AutoCAD has a comprehensive help facility. If the operator is uncertain of what is to be done next, then, by using a special button on the graphics tablet puck, a 'flip-screen' can be displayed. This contains a detailed description of the command's action and the parameters which must be input.

Systems provide various kinds of visual feedback, the cross-hair cursor being one example. In AutoCAD, a special area of the display is devoted to showing the current coordinate position of the cursor. Many systems use *rubber-banding* where a line is anchored at one end to the last point picked and at the other to the current position of the cursor. DOGS allows the user to change the drawing datum; the current datum being shown by a small cross. DUCTdraft permits the user to define several datum points and to choose one when needed during the drawing; all defined datums are displayed but the chosen one is shown larger than the others.

Another important user-aid is the screen layout. The display can be partitioned into sections which have a special purpose - we have already mentioned an example in the coordinate display used by AutoCAD. DUCTdraft uses an elaborate system of windows.

Elementary Draughting Operations

All draughting systems have drawing options which replace the actions performed in manual draughting. They consist, in the main, of procedures for defining points and drawing straight lines, arcs, circles and other curves. These components of a drawing are sometimes known as *entities,* each having its own definition. In some systems, and AutoCAD is an example, entities can be grouped together and treated as a whole. These are called *blocks.* Some of the commoner entities and the actions that can be performed on them will now be discussed.

Points

During the course of drawing, users need frequently to refer to points on features that have already been drawn - the intersection point of two lines is an example. Even if the user had the steadiness of hand and keenness of eye exactly to locate and pick a point with the cursor, it would be neither convenient nor fast. In practice, it is possible to pick a point or other entity if the cursor is within a certain distance from it. The pick 'snaps' to the entity.

On DOGS, a tolerance circle is displayed at the bottom left-hand corner of the drawing area; if a suitable entity lies within the radius of this circle centred at the current cursor pick position, then that entity is picked. If there is more than one suitable entity, then the first one found in a search through the drawing structure is taken. To pick one from a cluster of similar entities is sometimes difficult, or impossible. It is then necessary to zoom in on that section of the drawing to give more separation between the similar entities. The tolerance circle can be switched off at will and pick coordinates are then taken literally from the cursor position.

In AutoCAD and DUCTdraft elaborate snap procedures are provided - the user can snap to the centres of circles, centre points of lines and similar features. Snap grids are also provided; these are arrays of points at user-defined pitch which can be snapped to. This permits drawings to be done very fast without input of coordinates, if drawing dimensions are multiples of the pitch of the snap grid (which can usually be arranged). AutoCAD and DUCTdraft have two forms of snap grid; one is a rectangular array of snap points, the other, used less often, is a grid which can be used to draw isometric projections quickly and accurately. Besides picking points with the cursor, coordinates can be entered from the keyboard either in absolute form (relative to the datum) or in relative form (relative to the last point defined).

Lines

The simplest way of defining a line is by quoting the coordinates of its endpoints. The coordinates of the first point are entered in absolute form, the second may be relative to the first in either rectangular Cartesian (X and Y) form or in polar (angle and distance) form. AutoCAD permits the user to choose absolute or relative form, DOGS only allows relative form. In practice, the latter is not very restrictive, since it is the most natural method. DOGS makes a distinction between continuous and separate lines, continuous lines being joined together. Systems also allow lines to be drawn which depend on previously defined features, for example, tangential to circles and parallel to other lines.

Arcs and Circles

Arcs can be defined in many ways. Two possibilities are: centre–startpoint–subtended angle; startpoint–endpoint–point on the circum-ference. Most systems have a good variety available. Circles can also be drawn using different definitions. Like lines, circles may be drawn without explicit definition of coordinates: DUCTdraft allows users to draw a circle which is tangential to three other circles - in this case, some user help is needed to pick the required case from the many available. This is not uncommon in drawing lines and circles using other entities; for example, there are four possibilities for a tangent to two circles. It is also possible to construct continuous curves made out of arcs.

Curves

Some systems, for example AutoCAD, have standard entities for common curves such as ellipses: others, for example DOGS, do not. This is probably not because PAFEC did not consider adding this facility, but because they made a conscious decision to exclude entities which could not be drawn by compasses and ruler manually. Ellipses can be constructed in DOGS by other means, and they are then approximated by a series of circular arcs much as they are on the drawing board. Some systems also allow the definition of more general curves such as Bézier curves (which will be discussed later).

Deletion

It is possible on all systems to delete items by picking them individually with the cursor. Deletion of the entities within an area is also provided; the delete window is usually rectangular and its bounds are defined by cursor picks, but DOGS permits deletion within a user-defined polygon. The treatment of entities not entirely within the delete window may, in DOGS, be varied. They may be ignored or truncated within the window. In AutoCAD and DUCTdraft deletion inside the window can be selective; for example, all entities except circles can be deleted. In both of these systems, it is possible to put items in a delete buffer; they can be selected one by one and deletion deferred until all the required items have been picked.

Most systems also permit the deletion of the last item drawn without any further specification; some allow this process to be continued right back to the beginning of the drawing in the unlikely eventuality of that being required.

Line Fonts

It is a matter of convention to use different types of line in engineering drawing. The dashed lines that are used for hidden detail and the chain dotted lines that are used for centre lines are two examples. Also, the clarity of drawings is improved by using different line thicknesses - for instance, thick lines may be used for the outline of the component and thinner ones for dimension lines. Draughting systems permit operators to draw in various predefined line styles and thicknesses and also to define their own line types, if needed. It is difficult to draw thick lines on a raster display, particularly where lines intersect and it is becoming more common to use colour to distinguish them. Of course, the display is only a coarse representation of the drawing; with most commercial plotters, there is no problem and thicker lines are drawn with thicker pens.

Text

Text and dimensioning are probably the two facilities that impress new users the most on draughting systems. Annotations can very rapidly be added to a drawing in text of various predefined fonts and sizes, and at different angles. Users can define their fonts, if they have the time. It is often possible to justify text left, right or centre and some systems, including DUCTdraft, can automatically scale text so that it fits neatly inside a given box.

Hatching

Users can hatch areas using system-defined patterns. The area to be hatched must be a closed boundary (although DOGS does permit some tolerance on closure). If the boundary is very simple, it can be traced automatically by the system. If it is complex, the user must manually pick out the entities defining the boundary. Multiple boundaries can be defined so that holes can be left unhatched. Although the majority of hatching can be done easily and much faster than could be done manually, the odd case presents difficulties and a fair experience of the quirks of the hatching algorithm is needed.

Dimensioning

This, like hatching, gives a considerable advantage in time over manual methods. Since all the dimensions of a shape are known precisely by the system, dimensioning is merely a case of indicating the features to be dimensioned and the type and location of the dimension lines. It is also necessary to control the places of decimals and the tolerances. There is one minor snag. The system does not know whether the linear dimension between two points, say, is to be taken as the direct distance between them or as the projected distance in some other direction, typically horizontal or vertical.

The three example systems solve this problem differently. In DOGS, the angle of the dimension line can be set by the operator, so if a horizontal dimension line is needed, the angle is set to zero degrees. In AutoCAD, horizontal and vertical dimensioning are separate menu options. DUCTdraft displays rather more intelligence, the position selected for the dimension line being taken as a hint to the angle of dimension line needed. All respectable systems provide a good selection of dimension types, diametral and radial dimensions, and chained dimensions to a fixed datum line are examples.

Zoom and Pan

Since the screen display is only a comparatively crude representation of the drawing because of resolution limitations, it is usual for operators to work on magnified small sections of the drawing. This is made possible by zoom and pan facilities. The section of the display to be magnified is defined, often with the corners of a rectangular window being picked with the cursor. The section so defined is magnified to as large a size as is possible on the screen. Skilled users work on a drawing view by view in this way, but novices tend to work on the entire drawing since they are used to viewing the whole of the drawing on the drawing board. This often leads to difficulties.

The facilities that have just been described mirror the procedures used in manual draughting and so it is possible to use them to do any drawing that can be done using traditional methods. It is unlikely, though, if the facilities offered by draughting systems were restricted to these elementary operations, that drawing throughput could be improved at all. It is in the more advanced drawing facilities that will be described in the next section that the potential for improving drawing productivity lies.

Advanced Drawing Facilities

All systems offer operations that have no direct manual equivalent. Since they vary considerably from system to system it is impossible to give a comprehensive list, but some of the more common will be described. Drawing manipulation facilities are specifically designed to speed up the draughting process and they take advantage of the orderliness that is found in the majority of components which have often some degree of symmetry and contain standard parts.

Dragging

It is frequently necessary to reposition parts of drawings. The most common reason is to create a well-laid-out drawing, but there are other circumstances in which the ability to *drag* sections of the drawing is useful, for example in nesting applications in sheet metalwork. The details of the dragging facility vary. DOGS has a wide variety of drag options, sections of drawings can be moved, with rotation, scaling and shearing. The user selects the part of the drawing to be dragged, either by defining a window around it or by picking the appropriate entities individually and holding them in a buffer, much as was done in the delete operation mentioned previously. In some systems, dragging is dynamic - the section to be dragged can be moved around the display as though it were adhering to the cursor and then planted at the desired location.

There is another type of drag available on some systems: a point can be dragged and all the lines of which it is the endpoint are also modified to maintain the relationship between the point and those lines. DUCTdraft has a generalization of this facility; the operator can set up *dependences* between entities so that a change to one of them has a side effect on the others and the original relationship is preserved. A simple case is that of two intersecting lines which are filleted. A dependence can be created so that, if one of the lines be dragged, the fillet is also moved and both lines retrimmed suitably.

Copying

This is an operation much like dragging except that the original version is maintained in position. Again, there are many variants possible: drawn objects can be mirrored about a line and drawn repeatedly with rotation about a point. Usually, when vendors are demonstrating their wares, they make great use of the copy facility which can achieve spectacular speed for a well-chosen shape. An example is shown in Figure 3.3.

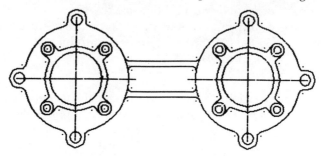

Figure 3.3   *Draughting system 'party piece'*

Layers

A drawing done on a draughting system can be organized as a set of overlays, each holding part of the drawing. These are called *layers* (or, in the case of DOGS, *views*). The maximum number of layers allowed varies, DOGS permits 40, DUCTdraft 256 and AutoCAD an unlimited number. Layers may be referred to by number or may be given names describing their use. Many CAD installations have developed their own standard for layer utilization, for example:

Layer 0	Draughting sheet
Layers 1–5	Written notes describing the component
Layers 6–10	Construction lines
Layers 11–20	Outline of the component
Layers 21–25	Dimensions
Layers 26–27	Hidden detail
Layers 28–29	Hatching
Layers 30–35	General annotation
Layers 36–39	Parts lists

Of course, it is highly unlikely that even the most complex of components would use as many as forty layers, but it is good practice to have some set standards. Layers may be organized in various ways; DOGS permits layers to be of any size, each layer having its own scale, AutoCAD allocates specific line fonts to each layer and so on.

The main usefulness of layering a drawing is that layers may be selectively displayed; the user can only work on one layer at a time - the current layer - but other layers may be displayed. This is beneficial for various reasons:

1. Working on separate layers speeds up operations such as redraws, since just one aspect of a drawing can be dealt with at a time. Thus, if a user were working on the dimensioning of a drawing, it would be a waste of time to redraw the hatching repeatedly. Also operations such as deleting a line are done faster because only part of the drawing file need be examined to find the line closest to the pick.
2. Construction lines can be placed on one layer and turned off very quickly, so saving selective deletion.
3. The use of layers introduces a potential for generality into drawings. For instance, alternative dimensioning systems may be used, metric and imperial dimensioning may be used interchangeably. Functional, manufacturing and gauging parts of the dimensioning system may be conveniently isolated so as to give clear indications of the dimensional requirements to different departments. Alternative drawing standards may be used.
4. Particular layers may be selected for specialized applications. The dimension layer may, for example, be turned off when drawings are to be shown to customers or included in catalogues.
5. Plotting may be done in a rather more flexible way; a group of layers may be plotted using one set of pens, then a further group done with another set.

Symbols

All draughting systems have a facility for defining and storing commonly-used symbols; in AutoCAD, symbols are called 'blocks'. A symbol is named and kept in a file, from which it can be retrieved and placed on a drawing at a desired point. A good selection of symbols can speed up drawing considerably in applications such as the drawing of schematics, where standard symbols for electrical, electronic, hydraulic and pneumatic components may be provided for the user. In detail drawing, symbols such as BS 308 standard tolerance boxes, machining symbols and common components such as nuts and bolts are often standardized for an installation. Firms may define their standard drawing sheet frames and logos in this way.

DOGS has an elaborate system whereby various types of symbol perform differently on retrieval; symbols may be scaled, rotated and sheared. DOGS differentiates between 'symbols' which are treated as entities, and 'shapes' which are drawings and may be edited. DOGS symbols may be *public* and stored in a symbols file which is protected from the ordinary user who may only retrieve symbols, and not delete or add them, or they may be *private,* and may be defined and used by the ordinary operator.

Parametrics

The facility which has the most potential for increasing productivity is that of producing parameterized drawings. In principle, if a firm makes a restricted range of products which have broad similarity, then drawings may be parameterized and the draughting procedure reduced to a clerical exercise. This extreme case is, fortunately for draughtspersons, rare; it is common, however, for some of a firm's drawings to suit this kind of treatment.

Often, firms producing families of components do not bother to draw all the members of a family to scale; they produce data sheets showing the general form of the component and a table showing the variant dimensions - tooling aids catalogues are laid out in this way. The disadvantage of this approach is that if an assembly containing this type of component is to be drawn then the component must be drawn to scale. Also, there are often slight variations of form which cannot easily be treated in this way. Drawings of components of this kind lend themselves to parametrization. A parametric definition of a family of components is like a data sheet but one in which the drawing of a particular component can be produced precisely to scale. In the more advanced versions, a parametric program can be produced which can deal with variations of form.

The ordinary symbol facility can cope with some component size variability but it is usually inadequate for any but the plainest of forms. A simple example is in the drawing of a bolt. In a view looking directly on to the head, a normal symbol could be used because change of size does not affect the geometrical similarity of the form. In a side view, the relation between the unthreaded and the threaded lengths is not dependent on a simple scaling factor and so a symbol is not suitable. Using parametrics, differential form variations of this kind can easily be handled. Using more advanced parametric facilities, shapes can be geometrically different. On a range of standard circular flanges, we might want to select a variable number of fixing-bolt holes, the number being chosen from a limited range of preferred values. This would be impossible using normal symbols but perfectly feasible using a parametric program.

The simplest form of parametric is when the operator does a sample drawing and is given the option of labelling some of the form dimensions as being variable. These variable dimensions can subsequently be allocated values. DUCTdraft provides this easily used facility. Some systems offer a special parametric programming language in which users can describe the process of doing a drawing.

Parametric programming languages are interpreters and contain most of the facilities that one would expect in a more general programming language - input, output, arithmetic calculation, mathematical functions, loops, conditional statements, subroutines and file-handling. The major difference is that the programming language also contains commands for performing the drawing actions possible with the system. A whole drawing could be described in this way, the user merely providing the system with drawing information in response to prompts, but it is more usual to use the facility for just part of a drawing.

As with most advanced facilities, the character of the programming language provided varies widely from system to system. DOGS uses a language not unlike BASIC; AutoCAD uses AutoLISP which is completely different from any language that the average programmer will have met, but is very interesting and powerful.

Because of the length of time needed to code and debug a parametric program of any complexity, it is important to establish that the payback in reduced drawing time will be worthwhile.

Non-drawing Facilities

Most systems, particularly the larger ones, offer facilities that are only indirectly connected with drawing. Again, there is wide disparity in the services provided and some systems offer them only as optional extras which are separately costed. A sample follows.

Measure

It is often handy for a user to be able to find out common properties of the elements of a drawing. A measure facility may be provided which can be used to evaluate line lengths and angles, and the perimeters, areas and centroids of bounded sections of the drawing. A built-in calculator may also be provided with some means of saving the results of calculation so that they may be input as dimensions when drawing commands are being carried out.

Properties

DOGS is one of the systems which provide a properties facility. Symbols may be provided with attributes such as cost and weight and each time a symbol is retrieved and placed on a drawing, some running calculation is performed. This may be a simple count, or the total cost could be progressively calculated.

This is particularly useful for applications such as plant layout and the design of hydraulic systems. Valves can be allocated a cost and the system will calculate the total cost as the drawing is in progress. It is also possible to allocate to cable and piping a cost per unit length, so that, each time a particular type is selected and a run drawn, the system keeps track of the total length and cost. A schedule of components can be prepared and, if necessary, added to the drawing.

Diagnostics

On some systems, and DOGS is an example, a diagnostic file may be kept. This contains statistics of the commands used in a particular session so that any problems with the system can be easily located.

Archiving

Micro-based systems such as AutoCAD provide little beyond a draughting capability and no assistance is provided with drawing office housekeeping operations such as storage and control of drawing release numbers and issues. There are separately sourced programs available which will perform this kind of action, but generally only larger firms use them. Usually, a company will have instead a modified form of their previous manual system of drawing storage and control. On more expensive systems, an archiving system is either provided as an integral part of the system or may be bought as an optional extra.

Plotting

On all systems, a drawing can be plotted and the plotting procedure is device-independent. Plot files may be created in some format which can be post-processed by the relevant plotter driver. It may be wondered why the drawing cannot be done directly from the drawing file. The main reason why a separate plot file is needed is that the drawing file is created from the actions taken by the operator and, since its purpose is to be displayed on a screen, no optimization is necessary. The speed of a pen plot, on the other hand, depends to a great extent on the order in which the separate lines are drawn; some attempt is made to optimize pen movement. This accounts for the rather eccentric movement of the pen when plotting. In a multi-user system, many operators share a plotter and it is usual to provide a plot-queue facility so that plot files may be entered in a queue and plotted when the plotter becomes available.

Intercommunication

Systems have their own unique formats for holding drawing files and these are usually of a fairly complicated structure so as to minimize entity search time. It is certain that if an attempt is made to transfer drawings from one proprietary system to another, they will not match. In order to communicate between systems without having to write translation programs, considerable effort has gone into the development of intermediate formats. IGES is one such format and it will be discussed later in some detail. AutoCAD's DFX format is also widely used. However, it is common for separate CADCAM subsystems to pass information from one to another, ideally through the medium of a database management system but typically through files. Some examples, using our three draughting systems, are:

DUCTdraft can pass drawing information to DUCT, a surface modeller and CAM system which will be described later.

DUCT can also pass finite element mesh files through a post-processor to finite element analysis programs.

AutoCAD can pass drawing information to PEPS, a popular micro-based CAM system, and a range of others.

DOGS can pass information to DOGSNC, PAFEC's own CAM system, and also to the PAFEC finite element analysis program.

Three-dimensional Modellers

Introduction

If we want to analyse some natural phenomenon, then we have to set up laws which describe its behaviour. As an example, we could describe a body moving down a slope by assuming that the slope is frictionless and that the only force acting on the body is that due to gravity (Figure 3.4a). This is a *model* of the real behaviour of the body. We know that slopes are not frictionless; so, if we want a closer approximation, we might consider the force system shown in Figure 3.4b. Again, if we are still dissatisfied with the closeness of this model to reality and have sufficient information about the body and enough technical knowledge, then we might take into account the air resistance (Figure 3.4c). At each stage the model has been *refined*, and we might carry on with the process until we ran out of theory, out of time or out of patience. Eventually, we might obtain a very close approximation to the real-life situation shown in Figure 3.4d, but it might well be that, for the purpose we have in mind, one of the coarser models would be quite adequate.

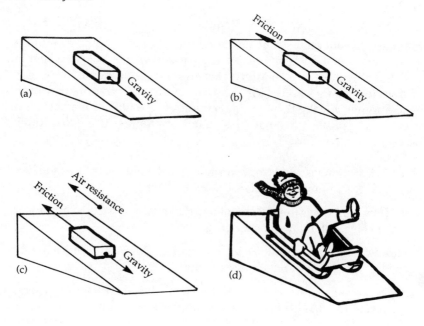

Figure 3.4   *Models of a body moving down a slope*

In a similar way, in CAD it is highly desirable to build and store computer models of engineering components. A very important quality of a component is its shape, which is modelled conventionally by the two-dimensional engineering drawing. There are, however, some advantages in holding a three-dimensional representation of a component. The main one is that the three-dimensional model is fully defined whereas the two-dimensional one need not be. As a bonus, designers can view the component from various angles, so that some idea of its shape can be obtained. Usually, engineers are content with just enough realism in the image to assess the suitability of the shape, unlike graphics designers who seek photographic (and better) realism in their images. There are other fringe benefits which are available to the engineer when a three-dimensional modeller is used; these depend on the type of modeller.

Consider the process of modelling a sphere. The computer has no knowledge of the form apart from the numbers that the user supplies to it. If a human were supplied with the coordinates of the points on a sphere of one metre diameter except for one point which appeared to be one kilometre from the centre, then they would probably apply their knowledge of the real world and engineering components and suspect that the point was in error. Computers currently are more gullible. It is, then, necessary to define the form as precisely as the application demands.

1. In Figure 3.5, the sphere is modelled by a mesh of lines. The computer model might be two lists, one of the space coordinates of each point and the other of the pairs of endpoints of the lines. Using this method, the system has very little information about the surface of the sphere. In particular, the same information would be held for a faceted body. A designer might guess that, since faceted bodies are not found commonly in engineering applications, the form intended was probably a sphere, but not so the system. If we wanted more surface detail, then it would be unable to supply it. This is characteristic of current systems; they have no 'world experience' and cannot deduce more information than is supplied. This may not be true of future systems.

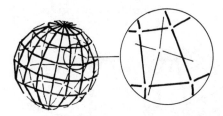

Figure 3.5   *Line model*

2. We might model the sphere by a set of flat panels - three-sided ones at the top and bottom, and four-sided ones elsewhere (Figure 3.6). The data to describe the sphere might be the same as in (1), but with lists of lines defining the panels. We have not gained much from this extra complication. Although the system could provide us with a shaded display which, if there were sufficient panels, would look smooth enough to be a sphere, it would not be justified in assuming that the model represented a sphere rather than a faceted figure. If part of the surface were to be machined, then it would be over-presumptuous of the system to produce a smooth, interpolated set of coordinates.

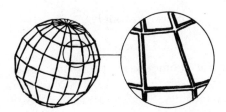

Figure 3.6   *Flat panel model*

3. Instead of flat panels, we could represent the surface by a series of

curved patches (Figure 3.7). These patches can be defined in a number of ways; the most popular are surfaces which, if sliced parallel with the three axes, give cubic curves - Bézier and B-spline patches are examples. Cubic curves are chosen because a cubic is the lowest order of curve which can, if desired, maintain smoothness at the junction with another patch. So, all the patches are defined mathematically, the coordinates on the whole surface are known precisely, and the surface is smooth. We cannot represent the surface of a sphere exactly using cubic patches, but we can be as close to it as is considered necessary for the application. There is another family of patches that will exactly model a sphere, but these have not the generality of cubic patches, which can be used to model virtually any useful surface closely. It is unnecessary that the surface should be smooth and we could model a cube or, for that matter, the faceted body in Figure 3.6.

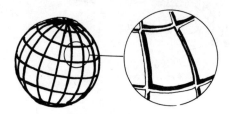

Figure 3.7   *Curved panel model*

4. A different approach might be to model a sphere directly. We might do this by defining an area - in this case a semicircle - and sweep it round a circle. Alternatively, we might define a selection of commonly-occurring bodies among which is a sphere. In either of these cases, not only have we defined the external surface, we have also defined the interior. The system deals with solids rather than with shells.

These four ways of modelling the sphere represent a series of increasingly refined models. They may be divided into three classifications:

Model 1 Wireframe model

Models 2,3 Surface model

Model 4 Solid model

We have modelled the geometric shape of the sphere but have not considered other attributes: its colour, its surface roughness, the density of the material from which it is made, and many of the other properties in which a designer might be interested. Although the geometrical form of the component is an important part of the database, other properties (its cost, for example) might be added if this were considered useful.

Examples of the three types of modelling system will be given in the sections that follow. The three systems are:

Wire-frame modeller   PAFEC's DOGS3D

Surface modeller   Deltacam's DUCT

Solid modeller   PAFEC's BOXER

Each of these will be used to model the simple body shown in Figure 3.8. Of course, with such an easy geometry, it is not possible really to put the systems through their paces, but the treatment is sufficient to give the flavour of three-dimensional modelling.

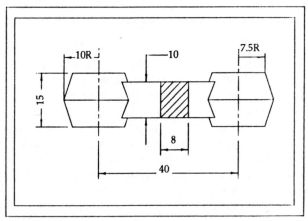

Figure 3.8   *The model*

Wire-frame Modelling

General Principles

The system to be described is a subset of the PAFEC DOGS computer-aided draughting system. To be fair to this system, it is rather more than a simple wire-frame modeller, since it permits some surface definition of forms as well as purely defining them by lines. It cannot, however, be described as a surface modeller because of the lack of generality of the surfaces that can be defined and PAFEC acknowledge this by offering a separate specialized surface modeller.

In DOGS3D, the display shows up to four elevations of the body at the same time. An example screen is shown in Figure 3.9. Positions of points in three dimensions are entered in various ways: by picking them with the crosshair cursor on two elevations, in Cartesian coordinates, in polar coordinates, in cylindrical coordinates and various other combinations.

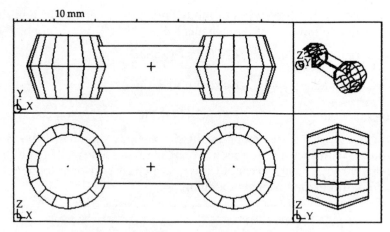

Figure 3.9   *DOGS3D screen layout*

The drawing options are similar to those described in the section on draughting systems; thus, we can draw lines and arcs and manipulate drawings by using copy, drag, delete. The only difference is that the drawing and manipulation operations are done in three dimensions - for instance, 'copy mirror' mirrors a volume about a plane. Another, more fundamental difference between this and a conventional draughting system is that the views displayed in the elevations are associative - changes to one view can result in changes to another if they are visible. Logical connection between views is not a feature of two-dimensional draughting systems and this quality is what makes the system a three-dimensional modeller.

A further feature of the system is that drawn lines and arcs can be automatically constrained to lie on the surfaces of planes, cylinders, cones and spheres.

Example

The following paragraphs describe the development of a model for the body shown in Figure 3.8.

A convenient datum is chosen. On this system, the datum may be moved at any time to suit the user's convenience. In this example, it will be fixed at the geometric centre of the body to be modelled.

The line joining points A at coordinates (20, 7.5, 7.5) and B at coordinates (20, 0, 10) is drawn. This is a generator line on the surface of the upper conic form making up the right boss.

The line AB is copy-rotated about the line $x = 20$, $z = 0$, which is the axial centre of the boss. The number of copies of AB is 15, equally spaced. The resulting set of lines describe the surface of the conic frustum (Figure 3.10).

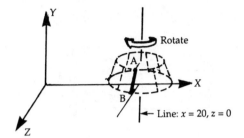

Figure 3.10   *Half the boss*

The circle at the top of the boss is drawn by picking the centre and existing points on its circumference. In order to pick a point with the cursor, it is necessary, in cases of ambiguity, to define the point completely by making picks on two separate elevations.

The lines making up the upper boundary of the bar joining the two bosses are drawn from the datum to the point where each intersects with the boss. The intersection line may be constructed, or its position calculated. In this easy case, the intersection points were constructed. In DOGS3D there is a facility called a 'fix' which can be used to force a point to lie on one of a range of surfaces; the cone is one of these surfaces and so the intersection point could have been obtained by supplying the y- and z-coordinates of the intersection point and defining the x-coordinate by using a fix to the conic surface of the boss. This facility is not typical of wire-frame modellers and so has not been used.

The upper quarter of the model has now been defined. This is mirrored about the plane $y = 0$ to give the right half of the body. The circle at the centre is added next. In the interests of clarity, the generator lines lying inside the volume of the central bar are deleted. Figure 3.11 shows the result.

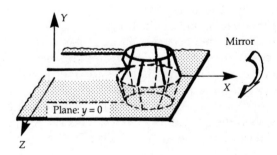

Figure 3.11   *The whole boss*

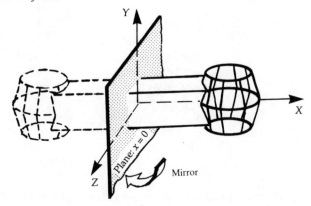

Figure 3.12   *The whole model*

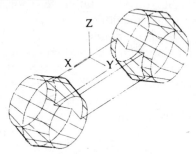

Figure 3.13   *Plot of wire-frame model*

All the model so far is mirrored about the plane $x = 0$ to give the complete model (Figure 3.12).

Superficially, it may not seem that there is much difference between the model produced by this modeller and that produced by the surface modeller that will be described in the next section. The main distinguishing feature is that we cannot automatically draw any intermediate surface lines on the wire-frame model - all that the system knows about the body being modelled is that the drawn lines correspond with lines on the surface of the body. The generator lines on the surface of the boss were not added by the system, but by the user.

As was previously mentioned, DOGS3D has a facility for defining surfaces - we could, for instance define the four frusta of cones as surfaces and we *could* then draw intermediate surface lines if we wished. This is not typical of wire-frame modellers which are usually quite primitive. DOGS3D also has the capability of 'unrolling' or developing shapes made out of planes, cylinders and cones, so that sheet metal patterns can be obtained. In developing the shape it takes into account the bending allowance and, to this extent, it can be regarded as a specialized system.

Surface Modelling

General Principles

Surface modellers are probably the most useful type of modelling system available currently. They are more sophisticated by far than wire-frame modellers and they are more versatile than solid modellers.

A popular surface modeller is Deltacam's DUCT, which will be used as an example. DUCT surface models are created in stages. The body is divided into pieces which are modelled separately and then combined. For most engineering artefacts, the division is obvious. For instance, in the shape shown in Figure 3.8, the two bosses are modelled separately from the centre section. A feature of surface (and solid) modellers is that the intersection points between the separate elements of such a body *can* be found automatically, in contrast to wire-frame modellers. Since all points on the surfaces are defined mathematically, points common to two surfaces can be calculated.

The surface of the body to be modelled is defined by a kind of lattice which supports the curved patches of which the surface is composed. Each patch is bounded by lines over which the user may have a high degree of control, but which can be given default forms, often perfectly satisfactorily, by the system. The surfaces can be smooth or discontinuous and are typically of a freer form than can be achieved with any other kind of modeller.

In DUCT, the surface can be altered interactively by the designer, which makes the system highly suitable for the design of 'aesthetically styled' products such as plastic detergent containers, telephone handsets and ceramic ware. In most forms, it is not necessary to perform the same kind of geometric simplification that is often necessary with solid modellers - we could very easily add blend radii and draft angles to the form shown in Figure 3.8. Blends can be added automatically by the system, which has other useful features that will be discussed later.

Example

The modelling of the form shown in Figure 3.8 will be described in the following paragraphs. DUCT, like most well-designed systems, allows the more experienced user to input data in a compact form and to take short cuts. In the interests of clarity, we shall not use any of these in the procedure below.

1. *Modelling the right-hand boss.*   First we define a plane section, then we rotate it to sweep out the boss surface. The section is defined in a two-dimensional Cartesian coordinate system in the plane of the screen. The horizontal axis is the U-axis, and the vertical axis is the V-axis. The section

is described by five points on its boundary, and the form of the lines joining the points may be controlled by the user. It is a common convention that when we are defining the boundaries of a shape by a line, then material is on the left of the line. DUCT uses this convention. Also, angles are defined using the standard mathematical convention of measurement from the horizontal with positive direction being anticlockwise. In this example, we shall start by just inputting the coordinates of the points and seeing what happens.

```
INPUT POINT 1   U       −10   V    −7.5
INPUT POINT 2   U      −2.5 V    −7.5
INPUT POINT 3   U        0   V     0
INPUT POINT 4   U      −2.5 V     7.5
INPUT POINT 5   U       −10   V     7.5
DRAW SECTION
```

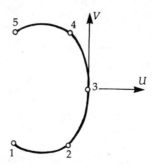

Figure 3.14   *Duct – first attempt at section*

The result is shown in Figure 3.14. That doesn't look too good! The system, lacking any other information, has done its best and presented us with a smooth curve through the five points, which was not what we wanted. Each of the lines that the system will place between the pairs of points will be a *Bézier curve* or parametric cubic. These are widely used in CAD systems because of their versatility; they can, for instance, be forced to straight lines which is necessary in this case. Point 3, for example, has a line entering it from point 2 and one leaving it to go on to point 4. DUCT allows the angles that these lines make on entry and exit from the point to be defined. An experienced user would have added these tangent angles to each point on input of the coordinates, but we have not been so far-sighted and so must edit the point definitions:

```
EDIT POINT 1   TAA   0
EDIT POINT 2   TAB   0           TAA   71.565
EDIT POINT 3   TAB   71.565      TAA   108.435
EDIT POINT 4   TAB   108.435     TAA   180
EDIT POINT 5   TAB   180
```

TAB means 'tangent before', TAA means 'tangent after'. The angles have been evaluated on a calculator, but the system would have been smart enough to recognize the expression '(ATAN(3.0))' which could have been inserted instead of '71.565'. There is a standard procedure for making a section out of straight lines, but it was considered more instructive to construct the section in the way shown. Let's see if that has made any difference:

DRAW SECTION

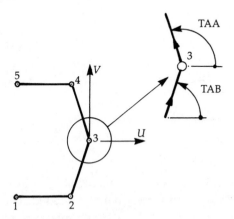

Figure 3.15   *Duct – corrected section*

The result, Figure 3.15, seems satisfactory. We shall need four of these sections, and so we shall store them for future use:

KEEP SECTION 1
KEEP SECTION 2
KEEP SECTION 3
KEEP SECTION 4

We have finished defining sections for a while. The next process is to define the curve on which the sections are to be placed. In this case, the curve is a circle and is called the *spine*. The spine is a curve in three dimensions, and DUCT uses the familiar right-handed Cartesian system. The circle will be placed on the X–Y plane and will be defined by four points, one for each of the sections that were stored in the previous paragraph:

INPUT SPINE 1	X	30	Y	0	Z	0	AZ	90	EL	0
INPUT SPINE 2	X	20	Y	10	Z	0	AZ	180	EL	0
INPUT SPINE 3	X	10	Y	0	Z	0	AZ	270	EL	0
INPUT SPINE 4	X	20	Y	−10	Z	0	AZ	0	EL	0
CLOSE SPINE										

The *X*, *Y* and *Z* coordinates should be self-explanatory. 'AZ' (which is an abbreviation for 'azimuth') and 'EL' (short for 'elevation') are two angles which define a direction in three dimensions (Figure 3.16). In this case they are used to define the tangent of the spine curve as it passes through each of the four points. It is also necessary to tell the system that the curve is to be closed or else there will be a gap between points 4 and 1.

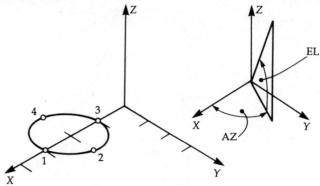

Figure 3.16   *Duct–spine definition*

We have now fully defined the boss and can examine it graphically (Figure 3.17). That seems to be what was needed. The four sections have been placed on the spine in the correct positions. DUCT will now add longitudinal lines around the section so that the surface is covered by a lattice of lines made up of the lines bounding the sections and the longitudinal lines. This lattice will be coated by a series of surface patches which are mathematically defined. The result is that every point on the surface is known to the system.

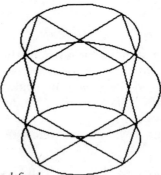

Figure 3.17   *The boss defined*

The resulting model is called a 'duct'. Figure 3.17 shows the basic duct with spine, sections and longitudinal lines; Figure 3.18 is a detailed duct, with some intermediate lines shown. The latter confirms that the model is a complete definition of the surface, unlike the wire-frame model in the previous section. The duct is satisfactory, so we can store it for future reference under the name 'RBOSS':

KEEP   DUCT   @RBOSS

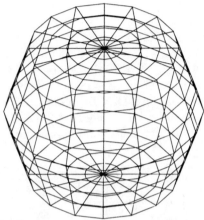

Figure 3.18   *Extra detail*

2. *Modelling the left-hand boss.*   Although the duct has been stored away, as @RBOSS, we still have a working copy of it which can be manipulated as needed. This is called the 'current duct'. If we move this 40 units in the negative *X*-direction, it can be used as a model of the left-hand boss, and stored away:

MOVE   DUCT   *X* −40
KEEP   DUCT   @LBOSS

3. *Modelling the bar.*   We shall hang two rectangular sections on a straight spine to define the centre section. Before any work is done on the sections, we must clear out the working store of the current section and duct.

DELETE   DUCT
DELETE   SECTION
INPUT   POINT 1   U   4   V   5 TAB 90 TAA 180
INPUT   POINT 2   U   −4   V   5 TAB 180 TAA 270
INPUT   POINT 3   U   −4   V   −5 TAB 270 TAA 0
INPUT   POINT 4   U   4   V   −5 TAB   0 TAA 90
CLOSE   SECTION
KEEP   SECTION 1
KEEP   SECTION 2

The spine must now be defined. It will be seen later that there is no need to bother about the exact point where the bar intersects the two bosses, since this can be handled automatically by the system. The bar will be modelled so that it falls short of the bosses.

```
INPUT SPINE    1  X  −10  Y  0  Z  0  AZ  0  EL 0
INPUT SPINE    2  X   10  Y  0  Z  0  AZ  0  EL 0
KEEP   DUCT       @BAR
```

4. *Modelling the rest of the bar.*   After presenting the system with a little more information, the extra bits between the end of the bar and the bosses can be obtained automatically with very little user action. The whole form has now been modelled and the resulting shape is shown in Figure 3.19. It will not have escaped the engineer's eyes that this is not a very good shape for a casting or moulding. DUCT could also have blended the bar to the bosses with a fixed-radius blend (Figure 3.20) or a free-form blend. We might also have added draft angles where necessary.

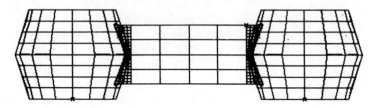

Figure 3.19   *The whole model*

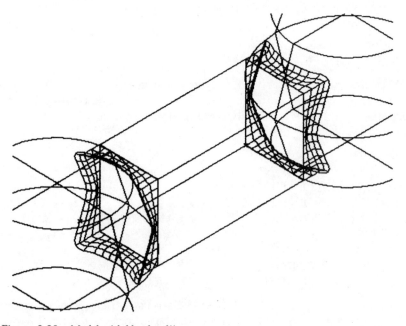

Figure 3.20   *Model with blend radii*

Further Operations

We have created a surface model of the component. This can be used as a geometric record of the part, but there are many other useful things that can be derived from the model. Some of these are given below.

1. We can view the model in three dimensions. The displayed picture can show the model from any viewpoint in the skeleton form that we have used previously. Although it is sometimes difficult to visualize the solid shape from this kind of display, it has the merit that it is displayed very quickly. The model can be displayed in isometric, perspective and orthogonal projections. It can also be shown as a stereoscopic pair and viewed through an inexpensive commercial stereoscope. It is more visually attractive, though far more expensive in terms of computer time, to display a shaded model. DUCT does not attempt to produce the ultra-realism of some commercial graphics systems but, given a suitable colour display, a reasonable image can be produced in any colour, with shading and specular reflections from multiple light sources, with various surface textures and transparency. The image is adequate for a designer to obtain a good idea of the form of a component and make decisions about its aesthetic appeal.
2. Surface areas and volumes can be obtained. Volumes are particularly useful for products like bottles, since the designer can work unrestricted by size considerations, and then, when the shape is satisfactory, can find the volume and scale the model accordingly. Shapes can also be developed if they are capable of being opened out to form a flat surface.
3. If the component is to be moulded, then part-lines can be found and draft angles added, where appropriate. The model that we have produced could do with a draft angle along the centre of the bar, for instance. The model was a solid form; the system can produce a cavity model from it. If the component is to be moulded, then the model can be passed to a mould analysis program so that its moulding characteristics can be evaluated.
4. A simulated machining process can be carried out on the model or the corresponding cavity. If the machining process is satisfactory, then a cutfile in a neutral format can be produced. This can be post-processed for input to a numerically-controlled machine tool.
5. The model can be passed to other CAD programs such as draughting systems like DUCTdraft and DOGS. It may then be dimensioned in accordance with normal engineering practice. This gives designers the capability of generating a component in a fairly free fashion and then defining it in a more rigorous way. This is a most useful way of working; many common artefacts are designed the other way round, their form being restricted by the draughtsperson's compasses and

ruler, and, worst of all, by the need to start with three orthogonal views. It is also possible to generate a finite element mesh from the model. This may then be passed to a finite element analysis system.

Solid Modelling

General Principles

The sample system will be BOXER which is marketed by PAFEC. BOXER is an example of a constructive solid geometry (or CSG) modeller; the model is constructed piecewise from a set of solids which can be manipulated and combined in various ways. The set of solids that BOXER provides are (see Figure 3.21):

Block (x,y,z)
Cylinder (l,r)
Sphere (r)
Cone (l,r,R)
Torus (r,R)

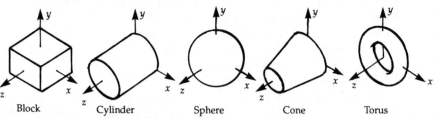

| Block | Cylinder | Sphere | Cone | Torus |

Figure 3.21   *Boxer primitives*

These basic building blocks are called *primitives* and we can create *instances* of them in order to build up the body. For example, the body shown in Figure 3.8 will be built up from four frusta of cones and one block. The primitives are defined in standard positions in a set of Cartesian coordinates but instances of them can be moved and rotated to any desired position. The instances can then be given variable names - in line with popular computer languages, we say that they have been assigned to a variable. In BASIC, we can assign integers, real numbers or strings of characters to variables; in BOXER, we assign solids.

As an example, suppose that we wished to create an instance of the primitive 'block', measuring 10 units in the x-direction, 20 units in the y-direction and 30 units in the z-direction, centred at the point $(x = 5, y = 6, z = 7)$, then we could do it by the assignment:

Chunk $<-$ BLOCK (10, 20, 30) AT (5, 6, 7)

The word 'Chunk' is the variable name which has been defined by the user, and is usually descriptive. The symbols '$<-$' together make up the assignment operator just as '=' is used in BASIC and ':=' in Pascal. The bracketed expression after the word 'AT' positions the instance at the required position. If we wanted to create another instance of a similar block, but rotated about the z-axis by 45 degrees, then an appropriate assignment might be:

> Chunk2 <− BLOCK (10, 20, 30) AT (ROTZ = 45, MOVEX = 5, MOVEY = 6, MOVEZ = 7)

We can combine solid variables using logical operators (Figure 3.22). In BOXER, the operations allowed are:

Union	+
Subtraction	−
Intersection	*

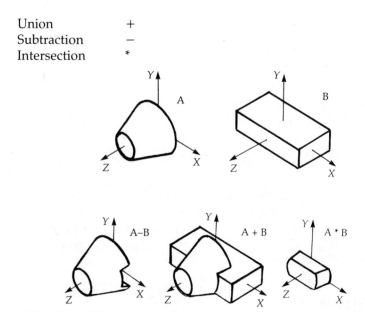

Figure 3.22   *Boxer solid operators*

Figure 3.22 shows examples of the use of the operators. Again, there is no unique way to model a body; if we wished to model a sphere with flats at two ends, then either of the two procedures below would be correct:

(a)
> Whole__sphere <− SPHERE (10)
> BlockA <− BLOCK (20, 10, 20) AT (MOVEY = 14)
> BlockB <− BLOCK (20, 10, 20) AT (MOVEY = −14)
> Flat__sphere <− Whole__sphere − BlockA − Block

(b)
> Whole__sphere <− SPHERE (10)
> BlockC <− BLOCK (30, 18, 30)
> Flat__sphere <− Whole__sphere * BlockC

The union operator joins two solids to make a whole. Often, we wish to model two touching solids, but wish them to keep their own identity. There is an 'assembly' operator, '&', which will do this. For instance, if we were modelling a box and lid, creating their union would result in one solid with an internal cavity. Using the assembly operation would allow them to be shown separately if we took a section.

Example

1. *Modelling the right-hand boss.* The boss is modelled by creating two instances of the primitive 'cone' and then gluing them together by using the union operator:

    cone1 <− CONE (7.5, 7.5, 10) AT (ROTX = 90, MOVEX = 20)
    cone2 <− CONE (7.5, 7.5, 10) AT (ROTX = −90, MOVEX = 20)
    right__boss <− cone1 + cone2

2. *Modelling the left-hand boss.* A copy of right__boss is created at the correct position.

    left__boss <− MOVE (right__boss) BY (MOVEX = −40)

3. *Modelling the bar.* The section joining the two bosses is created by creating an instance of the primitive 'BLOCK'. Notice that the length in the x-direction need not be exact and has been chosen large enough to fill the space between the two bosses without leaving any gaps.

    bar <− BLOCK (30, 10, 8)

4. *Modelling the whole component.* All that is required is to glue the component together, again using the union operator. It does not affect the model that some material is overlapping, since the solid union operator acts in an analogous way to the normal logical 'or' - if we take the union of two solids A and B, the result is another solid containing the material that is just in A or is just in B or is in both.

    whole__component <− cone1 + cone2 + bar

Figure 3.23 shows the complete model displayed with hidden lines removed and with generator lines added automatically to ease interpretation.

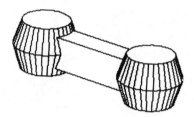

Figure 3.23   *Boxer solid model*

5. *A bit extra.*   The above procedure was so easy that we shall complicate the component a little by adding two holes of 5 units radius through each of the bosses.

hole__1 <− CYLINDER (20, 5) AT (MOVEZ = −10, ROTX = 90, MOVEX = 20)
hole__2 <− MOVE (hole__1) BY (MOVEX = −40)
total__component <− whole__component − hole__1 − hole__2

In the creation of hole__1, the order of the move and rotate operations is important. Another point to notice is that we could not have subtracted the holes before we added the bar or else the bar would have poked through the holes (Figure 3.24).

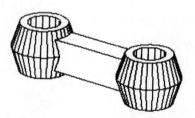

Figure 3.24   *Boxer – holes added*

Further Operations

We have created a model; what can be done with it? Some of the possible options are shown below.

We can draw it in isometric or perspective projections from any viewpoint. Figure 3.25 shows a perspective view. Notice that this is in accurate three-point perspective and so the bosses seem to be leaning outwards. An artist would probably draw the form in two-point perspective.

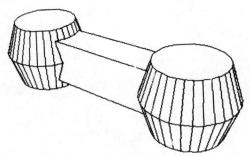

Figure 3.25   *Boxer – a perspective view*

We can examine it in orthographic projection (Figure 3.26).

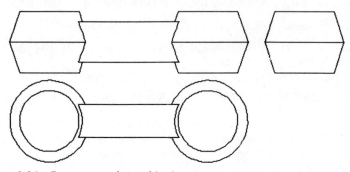

Figure 3.26   *Boxer – an orthographic view*

We can calculate its mass properties such as volume, position of centroid and moments of inertia.

We can take sections through it. Figure 3.27 shows a section which is at an angle to the centre-line.

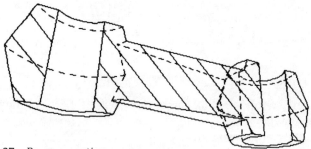

Figure 3.27   *Boxer – a section*

We can model neighbouring parts, then take intersections and so detect collisions.

Finite element analysis

We have discussed several models for the geometry of engineering components which are useful not just as a record of the shape of the part, but which can also be used to check functional requirements - whether the part is too heavy, whether its weight is distributed correctly, whether it fits inside the design envelope and so on. Designers are also interested in other characteristics of a component; for instance:

Whether it is stiff enough - or too stiff;
Whether it is strong enough - or over-designed;
Whether it will stand up to the working temperatures.

Very often, engineering shapes are not regular enough to permit analysis by the simple means taught on general engineering courses. Consider the beam model shown in the upper part of Figure 3.28. Structures consisting of straight lines loaded with arrows are not often found in practice and, in real life, the beam might look more like that shown in the lower part of the figure. Although we could use the crude model to find bending moments and shear forces, when it came to realizing the design in metal, the theory available to most engineers would not be sophisticated enough to find the local stress distribution around the pivot holes. In order to check that a design was satisfactory, we might make a model and use strain-gauges or some other experimental means. This is a sound way to validate the strength of the part, but it can take a long time. If the initial trial design is wildly out, several iterations must be carried out in order to achieve a satisfactory result and, if there is insufficient time to carry out these iterations, it is common practice to use large safety factors which can result in gross over-design.

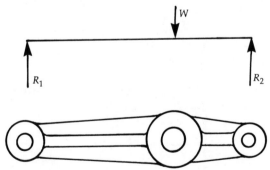

Figure 3.28   *Beam model*

An alternative method, now widely used, is finite element analysis (FEA). It is impossible to present a detailed description of this method here; it is explained in depth in many specialized texts. However, a non-mathematical outline will be given.

Figure 3.29 shows a flat component in tension. We shall consider this as being a plane stress system in that all the stresses are in the plane of the paper. If we rule lines on the surface of the part, and apply the loads, the lines will deform, something like Figure 3.30. In FEA, the part would be modelled by being divided into elements, possibly like those with boundaries defined by the lines that are drawn in Figure 3.29. If, by some mathematical wizardry, we could make the boundaries of the elements behave under load as the lines do, we would have no trouble finding the stresses. Unfortunately, in order to do this, we would have to know the answer before we started. What we can do, however, is to construct the elements so that they behave *approximately* like the shapes bounded by the lines.

We can do this by defining a set of points (or nodes) on the boundaries (and possibly the inside) of the elements and restricting the action of the nodes in some way that models the original body. We could, for instance, make sure that the displacements of common nodes in neighbouring elements are the same. Not only could we force continuity of displacement at nodal points, we might also force rotational, stress or strain continuity, each refinement of the model resulting in more complication in the subsequent analysis. It is usual to define the node constraint by assigning *degrees of freedom* to the nodes. The nodes on the elements shown on Figure 3.31, if constrained to give displacement continuity, would be described as having two degrees of freedom, since a displacement in two dimensions can be evaluated by two components - say, u in the x-direction and v in the y-direction in the figure.

Figure 3.29   *Component in tension*

Figure 3.30   *Deformed component*

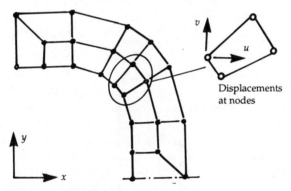

Figure 3.31   *Elements and nodes*

A *displacement function* is devised to describe the displacements within the elements. This is typically a polynomial in x and y. Using the principle of virtual work, the system sets up a *stiffness matrix* which connects forces at the nodes with displacements at the nodes, much as we connect loads and displacements of a spring by using the familiar formula

load = stiffness × deflection

We do know some things about the loaded component - the material properties, the loading conditions and the displacements at the constraints - and these are supplied to the system. The resultant model is a matrix equation or, equivalently, a set of simultaneous equations, which the system solves to give the required answers. A realistic model usually results in a large set of equations, which it is not practicable to solve manually; for three-dimensional models of any complexity, it is often necessary to use a large number-crunching mainframe.

Example

The user must first decide upon the design parameters:

the geometry
the material
the loading conditions
the constraints

The model can often be simplified, with resultant saving of processing time, by taking advantage of symmetry in the component geometry. In the plate shown in Figure 3.32, it is necessary to model only a quarter of the form, since there is no resultant displacement across the axes of symmetry. The other parts of the plate can be replaced by constraints. Another simplification can also often be made. Except in the region of the hole in the plate, the stress will be constant across the section, so that only the region of the hole need be modelled. The extent of the part that needs to be modelled is largely a matter of experience; in the case shown, it would be satisfactory to model the part shown in Figure 3.33.

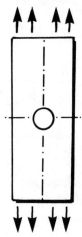

Figure 3.32   *Whole component*

Figure 3.33   *Portion modelled*

Next, the component can be divided into elements. The number of elements is again a matter of experience and a compromise is made between few elements and an over-crude model with very approximate answers, and many elements with excessive processing time. In this, as in many similar modelling systems, refinement of the model does not necessarily lead to more accurate answers after a certain point, because of rounding errors. Commercial FEA systems provide users with a comprehensive battery of different elements for different purposes - a few of those available in PAFEC 75 are shown in Figure 3.34.

If the results are to be reasonably accurate, it is important that the shapes of the elements should be chosen well. The closer a triangular element is to an equilateral triangle, the more well-behaved will be the model. The closer a four-sided element is to a square, the better. It is advisable to avoid large obtuse angles and large ratios between the longest and shortest sides of an element. A coarse mesh of reasonably shaped four-sided elements is shown in Figure 3.31. An obvious difficulty in preparing the mesh is that the elements are defined by the nodal coordinates. This can involve a lot of calculation which is not only time-consuming but is also prone to error. In order to reduce this overhead, either of two approaches is possible.

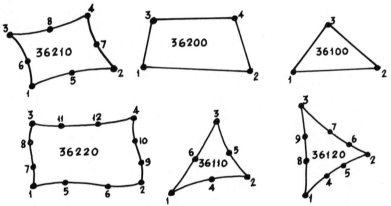

Figure 3.34   *Standard PAFEC 75 elements*

1. *Macroblocks* may be used. These are 'super-elements' which automatically generate a mesh within their boundaries. The spacing of the elements may be varied to suit the problem, a coarse mesh being sufficient where stress is fairly constant, and a fine one where stress has a high variation. PAFEC supply a set of macroblocks which they call 'Pafblocks'.
2. A *mesh-generating program* may be used. This is a front-end to the FEA system which is specialized to the generation and display of meshes and which may be used interactively. PAFEC provide a system called PIGS, which is provided separately from the FEA package and which considerably reduces the drudgery of calculating nodal coordinates.

The properties of the materials of which the component is composed may be already held in the system. PAFEC 75 provides sets of properties for ten common materials such as mild steel and concrete, each having a code number.

The user, having defined the model, can now supply the data to the system in an appropriate format. Data will include:

> Nodal coordinates
> Definition of elements
> Material properties
> Loading conditions
> Restraints

The data is usually held in a file, which has been entered using the standard editor, or has been prepared interactively using a mesh-generation program.

The system must also be provided with driving commands which determine how the job is to be run. These are specific to the problem, and to the system. PAFEC 75, like many other CAD programs, is highly modular. It consists of ten separate programs, which may be run in turn to give a comprehensive analysis. The stages of solution are:

1. READ. Data is input and stored on backing store.
2. PAFBLOCKS. The macroblocks are expanded to a full mesh.
3. IN-DRAW STRUCTURE. The defined component is drawn.
4. PRE-SOLUTION HOUSEKEEPING.
5. IN-DRAW CONSTRAINTS. As (3) but constraints added.
6. ELEMENTS. The stiffness matrix is derived and stored.
7. SOLUTIONS. Equations are solved, e.g. nodal displacements.
8. OUT-DRAW DISPLACEMENTS. Nodal displacements are drawn.
9. STRESS. Stresses are calculated from displacements.
10. OUT-DRAW. Stresses drawn in various forms, e.g. contours.

It is not necessary to run all the modules. For example, it might be considered uneconomic to produce a drawing of the component. However, this sort of output is very useful for detecting errors in input and model definition. Most systems can produce an 'exploded' drawing where the elements are shown with a gap between their boundaries so that holes in the structure due to omitted elements can be detected, which would be impossible if the elements were shown as defined. Examples of output are shown in Figure 3.35.

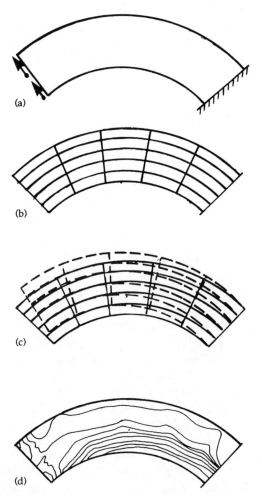

Figure 3.35   *(a) Loaded component (b) Model (c) Displaced shape (d) Maximum principal stress contours*

Finite element analysis of stress and strain is the numerical solution of a differential equation. It can also be used for many other applications, including:

Nonlinear static analysis - creep, plasticity, large displacements
Thermal analysis
Dynamic and vibration analysis

.It must not be assumed that the finite element method can be used in an undiscriminating way. Although the ordinary user need know very little about the internal workings of the system, in order to use it effectively considerable expertise is needed in the efficient formulation of meshes and in confirming the accuracy of the results. Despite the difficulties that arise in some applications, it is one of the most powerful and useful aids available to the designer and it may well be that future systems may incorporate artificial intelligence so that complex components can be analysed with ease by designers with only a modest knowledge of strength of materials.

Database Management Systems

'Plugging Away'

Scenario

E. Lecky and Sons is a firm specializing in the manufacture of electrical components. The Head Production Controller and his Senior Clerk are discussing a memo that he has recently received.

INTERNAL MEMO

From: The Chief Engineer
To: Head of Production Control

Brass Inserts in Plastic Plug Casings

Research and Development have been investigating a possible cost reduction in our range of standard domestic 3-pin plugs. It is proposed that the present threaded brass insert moulded into the casing should be eliminated since the new grade of plastic that we now use is strong enough to take a coarse thread. The current fixing screw will be replaced by a self-tapper which will reduce the cost by an estimated 1.2p per plug. I wish to explore the possibility of modifying other plugs in our range in the same way. I appreciate that you do not hold all the information, but I should like you to supply me with the following data as soon as you can:

(a) The part-numbers of our other connectors which are suitable for the modification. According to R & D, it is necessary for the thread length to be at least 5 mm in order to get a satisfactory clamping force and so it is important that the wall thickness should not be less than this figure.

(b) The number of these suitable connectors which are held in inventory. This will enable us to plan the phasing-out of the old stock.

(c) The probable savings over the next financial period, assuming that sales are the same as in the last period.

(d) The scrap value of the unthreaded brass inserts that we currently hold in stock.

Sorry to drop this on to you - if there are any problems in getting the information, please let me know.

HC: I don't need to tell you that this is a top priority job. Obviously, we don't hold all the information in the department and it will mean a lot of ferreting around. I'd like you to drop everything for a while and sort it out.

SC: I'm not quite sure how we can go about it. Have you any suggestions?

HC: Let's have a think. As a first step, you could go through all the assembly part-lists to check for threaded inserts. You can look through our card indexes.

SC: The Master Index should be the most reliable, but I shall have to check the amendment notes which are outstanding. Well, that shouldn't take too long. What about the wall thickness, though?

HC: Better get some help from the Drawing Office. They can check all the moulding drawings if you can give them the list of part-numbers that you've made up.

SC: That could take a long while. Still, they have plenty of time on their hands, don't they? The number of plugs in stock should be available from Stock Control; they should know the number of threaded inserts as well. It will probably be pure fiction, though.

HC: What about the weight of the unthreaded inserts? We'll need that to find out the scrap value. Don't the Drawing Office put estimated weights on the drawings?

SC: They used to, but I haven't noticed any on newer components recently. It will be easier and more accurate to weigh the actual components.

HC: That only leaves the sales figures. Sales are usually a bit cagey about letting their statistics out of the Sales Department. Probably they don't trust them. Leave that job to me. I'll lean on them a bit. Right then, I'll leave you to get on with it. Keep me posted on your progress.

SC: [*Leaves, muttering something under his breath.*]

■   ■   ■

The worst problems that Lecky's Production Control Clerk faces in collecting the information for the Chief Engineer are:

(1) the information is scattered around the company
(2) the information may be inaccurate or not up to date;
(3) some of the information may be duplicated
(4) some information is hard to find;
(5) most of the information is not directly available

Earlier, we mentioned the need for managers to have access to good quality information and to have access rapidly and conveniently. In Lecky's, as in most other firms, the information exists but is spread around the firm in different forms and so is not easily collected. Not only is this time-consuming but it also has an effect on quality since it leads to difficulty in maintaining accuracy. The problem is worsened by different departments having their own version of the same data - which not only wastes resources but makes consistency hard to ensure.

In order to guarantee the quality, ease of access and security of information, it is necessary to control it (and possibly hold it) centrally. An efficient way of doing this is to use a computerized information management system.

What is Required of a Management Information System?

1. Data should be *shareable*. All interested users who are allowed to obtain data should be able to do so. This does not necessarily mean that there should be a purely centralized 'databank' since, with computer networks, one workstation may often communicate with another. It may suit a firm's needs to distribute data in the sites where it is most used. This improves flexibility, permits easy expansion and has other advantages.

2. Data should be *non-redundant*. Traditionally, even with computer systems, files of data overlapped. Thus, an employee's name, clock number and job description might be held by the accounts department and separately by the personnel department. If an employee changed her name, it would be easy for one of the two files to be overlooked when the amendment was made.

   Even if there were some mechanism to ensure that amendments were made to all files containing the information, it might well happen that they were amended at different times so that data was inconsistent. Similarly, product description information might be held in the drawing office, in the cost office and other centres. In some manufacturing areas, changes to products are made very frequently and it is hard to make sure that a file contains the latest version of the product description.

3. Data should be *secure*. It is important that access to sensitive data should be restricted to persons who have authorization. The most

obvious reason for this is the danger of industrial espionage. New designs of motor cars are often kept shrouded until they are announced. There is a lot of non-technical information that is equally commercially confidential. However, the major reason for restricting access to privileged users is the danger of data being corrupted either by accident or maliciously.

4. Data should be *easily accessible*. This does not only mean that access to data should be convenient. It means also that users should be able to ask for data in a natural and flexible way. Very often we do not have the precise information that will enable us to find out what we want to know. The system should be capable of making reasoned guesses and giving us a list of alternative answers when we have asked an imprecise question.

   From a telephone directory, if we know the name, initials and address of the person that we wish to ring, we can easily find the number. If we only know that their name is Mudassar or Murphy, then we have problems. If we can only remember that their first name is Bridget and they live in Bond Street, Birmingham, then we *really* have problems. The system should ideally be able to cope with cases like these. It should also support 'browsing'.

5. Data should be *centrally administered and controlled*. Even if data is not held centrally but is distributed, it is usually better to control it centrally in order to ensure its quality and non-redundancy.

6. Data should be capable of being accessed rapidly . In any computerized system, there is always a trade-off between speed and size of storage needed. Suppose that there were suddenly a petrol shortage and the firm had to lay on coaches to pick workers up from home and take them back. It would be a tedious affair for a large firm to schedule the coaches using the personnel name and address file, because this would probably be ordered alphabetically by name - like a telephone directory. In order to speed things up, they would have to create another file sorted by district or post code. They would then have traded off storage space in order to improve speed. Because of this 'space–time compromise', it is true in general that rapid access involves large storage.

Features of a Database Management System

Lecky's problems would be solved if they used an integrated database management system (DBMS) with on-line access. Although the term 'database' is often used to describe a large file of any kind, it really means more than this. A DBMS has the qualities that we have just described; the mechanisms which provide these qualities will now be discussed.

A typical database contains data which may be used for many applications. If there is to be no redundancy and if access is to be controlled, programs cannot use the database directly. There exists a detailed definition of the format of the database and other relevant information. This is called a 'schema'. The schema contains the data definition, details of the links between the data items and information about user access permission.

Programs are supplied with a 'sub-schema' which defines how the program wishes to 'see' the section of the database with which it is concerned. The sub-schema acts as a kind of window through which an applications program can only see the data that is relevant and see it in a form which is appropriate for its needs. The linking and access control between the database and the program is done by the DBMS. This mechanism permits non-redundancy of data, ensures that data is independent of the program processing it and that access is controllable.

Users may be given various levels of privilege to retrieve, create, delete and update data. At the lowest level of privilege, they are permitted only to access data which is relevant to the application on which they are working. At the highest level of privilege, they are allowed all possible operations on all data. Clearly, great care must be exercised in granting such privilege because the setting up of a database takes a lot of time and is therefore costly.

Of course in any realistic system, there will be procedures for periodically backing up the database and, usually, there are provisions made for recovering data in case of corruption. Even with these safeguards, some of the most recent transactions may be lost. In firms with large database commitment, there is often a section separate from the computer department and run by a Database Manager, who is responsible for the control and granting of access rights.

Most data, and this applies especially to engineering data, displays a structure. Items of data do not exist in isolation: they have relationships with other data items. Consider a simple domestic electrical plug. If we wished to represent the assembly in a file, we could possibly just list a description of each of the nineteen items that make it up together with their part numbers:

Description	No	Part No
Top case	1	T362
Fibre clamp	1	T123
Screw	2	T064
Pin	2	T518
Screw	3	T452
Pin	1	T430
Spring clip	2	T124
Rivet	2	T023
Seat	1	T400
13 amp fuse	1	T423

Bottom case	1	T425
Screw	1	T800
Adhesive label	1	T912

This is called a 'flat file'. It is a compact method of representing the assembly but does not display the structure of the component. A more useful representation might be:

3-pin plug assembly	P124
Top case	T362
Cable clamp	P234
Fibre clamp	T123
Screw	T064
Screw	T064
Earth terminal	P568
Pin	T518
Screw	T452
Neutral terminal	P569
Pin	T518
Screw	T452
Line terminal	P634
Terminal pin	P635
Pin	T430
Spring clip	T124
Rivet	T023
Fuseholder	P636
Seat	T400
Spring clip	T124
Rivet	T023
Screw	T452
13 amp fuse	T423
Bottom case	T425
Fixing screw	T800
Adhesive label	T912

This is a 'structured' file and in many circumstances would be a much more useful representation since it carries information about the *structure* of the assembly as well as its components. It might well be that sub-assemblies are packed together and occur in more than one type of plug. In this case, it would be useful to just carry a reference to the sub-assembly in the parts list of the plug.

The example given above is very simple; more complex assemblies can consist of several tens of thousands of components. The files that we have created are also elementary. The part numbers might, in practice, be pointers to separate structures describing the geometry of the components - possibly a drawing file or, more usefully, a three-dimensional model. In the latter case, it would be possible to obtain, for instance, the weight of brass in the assembly. Other useful attributes would be pointers to standard material specifications, manufacturing information, tolerances and so on. Using this technique, a complete model of the assembly could be built up, with as much detail as was considered useful. There are various types of database management system but the normal user need know very little about the internal structure and working of their particular system.

Problems

1.   When working fast, manual draughtspersons will often use techniques which are either not available or not as convenient on computerized draughting systems. For instance, they often draw corner radii freehand. A possible reason for this is that input devices are too inflexible; another may be that system designers have not studied manual draughting sufficiently. A solution might be to have an input device which is more like a conventional drawing board, with a tee-square, ruler and adjustable square, and which can be switched between manual and computer operations, the manual operations being automatically digitized.

Write a short letter to the supplier of your draughting system advocating this.

2.   Professor Ben Trovato has come up with another brainwave: 'I don't know what all the fuss is about three-dimensional modelling. All you have to do is to allocate a sufficiently large block of store which represents three-dimensional space. Each storage location represents a small cubic volume in the modelling space and is marked with a binary one if it lies mainly within the component that you are modelling. It could also contain details of colour. Showing a shaded model would then be easy, removing hidden detail trivial and finding the mass would just be a matter of counting the locations which are marked.'

Assess the feasibility of the suggestion.

3.   A problem with the design of draughting systems is the user's expectation of response time. An engineering drawing can contain hundreds of lines which are stored internally in some data structure. To draw all the lines in the order in which they are stored might be quite a fast operation, whereas to search for the line closest to the cursor and delete it might take a considerably longer time. The users' expectations of these operations would probably be based on their experience with manual methods, where to repeat the drawing completely would take far longer than to delete a line.

Investigate how system designers solve this problem and at what cost.

4.   Properties of a component such as its geometric shape, its colour, its material and its weight can all be modelled in computer systems. But often there are other properties which affect our feelings towards the component - for instance, its feel and its smell.
Discuss the utility and feasibility of modelling these also.

5.   Professor Ben Trovato read the proofs for this book and disagrees with a statement in Chapter 3 - 'However, current systems suffer from some drawbacks. A given system can only model some of the possible shapes in which a designer may be interested.' He claims that 'This is not a drawback at all. It is an advantage. Surely that is what fashion and house styles do, anyway. If we select our primitives and permitted operations in a three-dimensional modelling system judiciously, then we can force the designer to conform to a desired fashion because fashion is largely a matter of shape. We might even be able to avoid some of the bizarre designs that we currently see around us.'
Discuss whether this point of view is valid, contains an element of truth or is just plain silly.

4 Elements of computer graphics

Introduction

In this chapter are considered some of the operations needed to display pictures on graphics screens. The discussion will be limited to raster displays, since these are the most popular for CAD applications. Computer graphics is a subject in its own right and is developing very rapidly, so only a broad treatment will be given; detailed mathematical analyses and computer programs will be found in specialized books on the subject. An introductory text that covers the subject thoroughly is M. E. Mortenson's *Computer Graphics: An Introduction to the Mathematics and Geometry* (Heinemann Professional Publishing/Industrial Press, New York).

Many of the graphics operations that we describe will be 'easy' - that is to say, a human being could perform them without much difficulty. The reason why the computer versions appear so complicated is that there is a constant demand for increase in the speed of graphics and any ingenuity in the procedures is mainly to do with making them as fast as possible. A major aim in graphics system design is to be able to present pictures that are animated in real time and to present them in a realistic way, shaded, with highlights, in colour, with texture and translucency and so on. It is probable that processing speed in the near future will be improved by using parallel processors, but today's systems designers struggle with more conventional means and it is vital that operations should be as fast as possible.

Elementary Operations

Scan Conversion

A raster screen can be thought of as a rectangular grid of dots upon which points, lines and curves can be represented by filling in selected dots. For the moment, we shall assume that the features to be drawn all lie within the boundaries of the display and that they are scaled to the actual size at which they are to be displayed in screen units. As an example, suppose that we wish to draw a line joining two points whose screen coordinates are known. We must decide which intermediate pixels are to be turned on. This must be done with a minimum of calculation, by integer arithmetic if possible and in a way that is suitable for implementing by hardware. The need for integer arithmetic arises because pixels are addressed as a pair of whole numbers (possibly coded) and computers work faster using integer arithmetic than they do using real numbers.

Processes like this are examples of what is known as *scan conversion* - we met the term before in the section on electrostatic plotters - and similar operations occur fairly often in graphics applications. A method, easily implemented in hardware, is Bresenham's algorithm. A section of the line to be drawn is shown in Figure 4.1. The problem is to choose the pixels around the theoretical line so that they give as good a representation of the line as possible. At the point shown in the enlarged section of the figure, it has to be decided whether to chose pixel A or pixel B. Clearly, the one chosen has to be the one lying closer to the true position - in the example shown, pixel A is the right choice. Bresenham's algorithm selects the correct pixel in a way that uses only integer addition and multiplication by 2 (which can be performed fast by a shift left by one place of the binary number) and so is highly suitable for implementation in hardware.

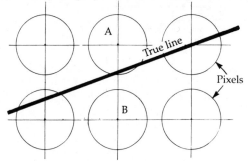

Figure 4.1   *Scan conversion – lines*

A simplified procedure is shown in the flowchart in Figure 4.2; the line variables X and Y are as shown in Figure 4.3 and, for convenience, the line is assumed to start at the bottom left-hand corner of the screen with a slope less than 45 degrees. It can easily be modified for other cases. For the example shown in Figure 4.3 the values obtained are:

$$X = 8, \quad Y = 3$$
$$e_1 = 6, \quad e_2 = -10$$
$$d = -2$$
$$u = 0, \quad v = 0$$

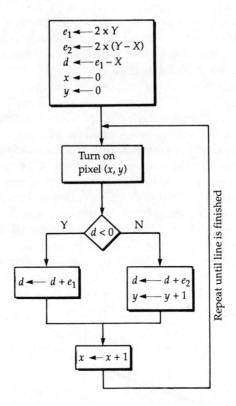

Figure 4.2   *Simplified form of Bresenham's algorithm*

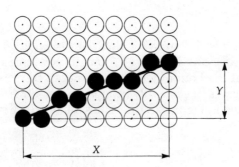

Figure 4.3   *Bresenham's algorithm*

x	y	d
0	0	−2
1	0	4
2	1	−6
3	1	0
4	2	−10
5	2	−4
6	2	2
7	3	−8
8	3	−

Similar methods are used for the scan conversion of circles and other mathematically defined curves. We shall not go into any more detail because even specialized graphics systems designers need never program the algorithm; the operation is built into the display logic in any but the most primitive of graphics displays. It has been included here to make the point that even drawing a straight line is not straightforward in graphics work.

Block Fill

Most modern raster screens provide a facility for filling in areas of the screen in one colour or, on some displays, in standard patterns. In shaded images, the tonal graduations are supplied by filling in the facets or panels that make up the surface of the body being modelled and, once the vertices of a facet have been calculated in screen coordinates, the process is fairly automatic.

A common way of filling in blocks is *flood filling*. Since a closed boundary on the screen has an area inside and one outside, the display logic must be provided with a *seed*, a point which lies either inside or outside the boundary, and so defines which of the two possible areas is to be filled. Pixels surrounding the seed are checked and if they are inside the boundary and inside the screen limits, they are filled in. This is continued while there are any eligible pixels (Figure 4.4). As in many processes in graphics, validation is sacrificed for speed and so, if there is a gap in the boundary, the fill pattern leaks through it and the fill may not

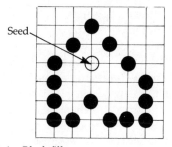
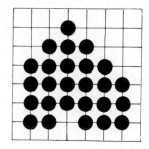

Figure 4.4   *Block fill*

be what was intended. Again, this procedure is often implemented in the hardware of the graphics display.

Clip and Cover

It often happens that we do not want to display all of a picture; for instance, when we zoom into a drawing, it is necessary to *clip* the picture so that no attempt is made to draw any part outside the limits of the screen. The portion, often rectangular, of the picture which is to be displayed is called a *window*. Again, occasionally, we wish to superimpose some message on a picture and it improves clarity if we blank out the part of the picture surrounding the text. The methods of eliminating unwanted sections of a picture from outside or (less commonly) inside a rectangular area are called *clipping and covering* algorithms. Since they are used very often in graphics work, like scan conversion and block filling, they are commonly implemented in the hardware of the display.

There are several methods for clipping - a popular one is the Cohen–Sutherland algorithm. A crude method would be to take each line on the picture and laboriously check where it intersected one or more of the four lines making up the rectangular boundary of the clipping window, then to discard line segments outside the window. This brute force approach would waste time. A lot of the lines might not need clipping because they lay completely *inside* the window and could be drawn unchanged, or completely *outside* it and could be ignored. The Cohen–Sutherland algorithm identifies these lines rapidly and so reduces the amount of calculation.

The algorithm firstly takes the rectangular window and extends the lines bounding it, as shown in Figure 4.5. This divides the display plane into nine regions. Each of the regions is allotted the four-bit binary code shown.

Bit 1 set   A point is left of the window
Bit 2 set   A point is right of the window
Bit 3 set   A point is below the window
Bit 4 set   A point is above the window

The binary codes corresponding with the endpoints of a line which is a candidate for clipping are calculated. If both codes are zero, then the line lies entirely inside the window, does not need clipping and so can be displayed in its entirety. If the bit-by-bit logical AND of the codes for both points is non-zero, then the line lies completely outside the window and so does not need to be displayed at all. Otherwise, the line is to be clipped, the points of intersection with one of the window lines are found and the appropriate part of the line is discarded. The process is repeated until what remains of the line is within the window. An example is shown in Figure 4.6.

1001	1000	1010
0001	0000	0010
0101	0100	0110

Figure 4.5   *Clipping: Cohen–Sutherland algorithm*

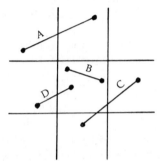

Figure 4.6   *Examples of clipping*

Line	*End Codes*		*AND*	*Result*
A	1001	1000	1000	Non-zero, so ignore
B	0000	0000		Both zero, so draw
C	0100	0010	0000	Zero, so clip
D	0001	0000	0000	Zero, so clip

This is typical of graphics algorithms in that it attempts to reduce the time taken by the crude brute force approach. The algorithm may be modified for covering.

Curves

The commonest geometrical forms used in engineering are lines and arcs of circles, probably because they are easily obtained by straight-edge and compasses. There are often theoretical reasons why fillets should not be circular: stress concentrations are lower on hyperbolic fillets. Yet circular fillets are almost always used. This restriction on form due to the limitations of manual draughting does not apply in most computer-aided systems. Using surface modellers it is possible to define freely forms that are more pleasing to the eye than those obtained by conventional manual means. The problem with defining free-form curves is that it is important that the designer should be able to arrive at a desired form by trial and error.

The internal representation of lines and arcs of circles is simple and compact; for instance, we can completely define a straight line by specifying the coordinates of its endpoints. The plotting of lines and arcs is often done using firmware built into the display logic, as we have seen. However, a method for representing curves of a more general form is not immediately obvious and has been the subject of much research. Probably the most natural solution is to define a series of points on the desired curve and then fit an appropriate polynomial to them. Thus, if we defined twenty points on a curve, we could calculate the twenty coefficients of the nineteenth order polynomial which would pass through the points.

This seemingly straightforward solution has serious drawbacks. For our twenty points, it would be necessary to solve a system of twenty simultaneous equations (or invert a 20 × 20 matrix which is equivalent). This might not be considered to be too prohibitive a task on a computer, but even the slightest change required in one of the points would necessitate a re-calculation of the coefficients: the method would not be suitable for interactive work where it is usual to arrive at a solution after many iterations. Another problem is that there is no guarantee that the curve between the points would be smooth. For example, if it were desired to fit a cubic polynomial to the four points (1.0, 3.0), (2.0, 2.9), (3.0, 3.0), (4.0, 7.0), it might be imagined that the curve would be simple, whereas the result would in actuality be as shown by the broken line in Figure 4.7.

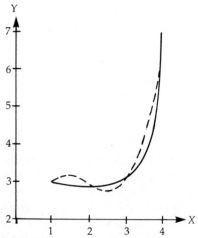

Figure 4.7 *Curve fitting*

A preferable method is to divide the curve into separate sections (sometimes called 'spans') and fit each section individually. Arrangements can be made to match the sections at their boundaries so that the total curve can, if wished, be smooth. One technique is to use 'splines', the properties of each section of the curve being chosen so that its end tangents and possibly curvatures are the same as those of its neighbours. The separate spans are defined by simple polynomials such as cubics or biquadratics.

One of the more popular techniques is to use Bézier curves (named after the French CAD pioneer who first developed them). A span is defined by four points, two being the endpoints of the span and the other two 'control points' which lie off the curve (Figure 4.8). The endpoints locate the curve, while the control points govern the slope of the curve at its ends and its 'bulginess'. The Bézier curve is a cubic, which is a good compromise between the need to provide a sufficiently wide range of possible forms between the two endpoints and the need to provide a sufficiently wide range of possible forms between the two endpoints and the need to have a compact representation. Using a cubic enables the curve to take a diversity of forms. It is not possible to define an arc of a circle *exactly* but by picking suitable control points, a very close approximation can be obtained.

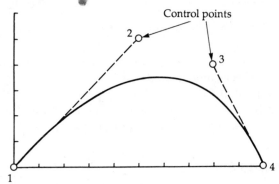

Figure 4.8 *Bézier curve*

To control the shape of a curve by manipulating points which do not lie on it does not sound easy but with a little practice it is possible for users to achieve a high degree of facility. It is easy to change the control points interactively and view the resulting effect on the graphics display; the user soon gets skilled at attaining a desired result. This is closer to the natural way that designers work than by forcing them to define separate points on a curve, and the method ensures that the curves are well behaved without any undesired inflections.

It is useful to imagine that the curve is bounded by a quadrilateral which has sides made up by joining the endpoints and the control points (Figure 4.9). A property of the Bézier curve is that it lies completely within this quadrilateral. Often, the two lines joining an endpoint and a control point are called 'control lines'. The angle of the control line determines the tangent of the curve at its ends; the length of the control line determines the shape of the curve - the longer the control line, the more the curve 'sticks' to it. It is usual to represent the cubic describing a curve in parametric form. If the coordinates of the endpoints and control points are numbered as in Figure 4.8, the parametric equations of the Bézier curve through them are:

$$x = x_1 (1-t)^3 + x_2 3t (1-t)^2 + x_3 3t^2 (1-t) + x_4 t^3$$
$$y = y_1 (1-t)^3 + y_2 3t (1-t)^2 + y_3 3t^2 (1-t) + y_4 t^3$$

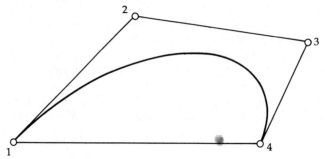

Figure 4.9   *Bézier curve – bounding quadrilateral*

The parameter t varies between 0 and 1. These equations can be extended to apply to the three-dimensional case.

As has been noted, Duct uses Bézier representation very effectively to produce easily surfaces that are visually pleasing. There are other methods commonly used, the most popular being B-splines.

Transformations

Introduction

The following treatment needs a knowledge of matrix addition and multiplication.

Point representation.   A point is conventionally represented in two dimensions by two coordinates $\{x\ y\}$ and in three dimensions by three coordinates $\{x\ y\ z\}$. It is, however, more convenient in graphics work to use *homogeneous coordinates*. In two dimensions, the point is represented by *three* coordinates $\{x\ y\ 1\}$ and in three dimensions by *four* coordinates $\{x\ y\ z\ 1\}$. These may be regarded as vectors or 3×1 and 4×1 matrices. The 'redundant' element is always 1. The curly brackets indicate that the vectors should really run down the page.

Line representation.   A straight line in either two or three dimensions may be defined by its endpoints.

Panel representation.   A panel (or polygon or facet) in two or three dimensions can be represented by the straight lines which make up its boundaries. The triangular panel in Figure 4.10 may accordingly be described as follows:

Points	1	(1 1 1 1)
	2	(4 5 2 1)
	3	(2 4 3 1)
Lines	1	(1 2)
	2	(2 3)
	3	(3 1)
Panels	1	(1 2 3)

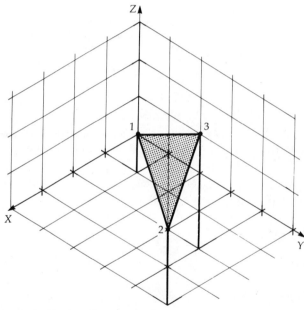

Figure 4.10   *Point/line/panel representation*

We can build up complex curved shapes in this way, by using polygonal panels which, if they are small enough, can approximate the surface very closely. This simple approach will serve us for the moment and is used in wire-frame modellers: in more advanced systems, the representation used is generally more complicated.

Two-dimensional Transformations

Often, we wish to move a drawing around the display area. When using a draughting system, we may wish to reposition a view or copy it to another part of the drawing. This sort of operation can be done by using combinations of three basic transformations:

Translation
Rotation
Scaling

Each of these can be represented, in two dimensions, by a 3 × 3 matrix. The point positions of the transformed image can be found by multiplying the coordinates of the points by the suitable transformation matrix. This unified approach is made possible by the use of homogeneous coordinates: if we used the conventional point representation, we would need a procedure for translation which would be different from that used for rotation and scaling.

Translation

If it is needed to translate (or move) a point in two dimensions by T_x in the x-direction and T_y in the y-direction, then the transformation matrix is

$$T = \begin{matrix} 1 & 0 & T_x \\ 0 & 1 & T_y \\ 0 & 0 & 1 \end{matrix}$$

Suppose we wanted to translate the triangle in Figure 4.11 by two units in the x-direction and by three units in the y-direction; the matrix would be:

$$T = \begin{matrix} 1 & 0 & 2 \\ 0 & 1 & 3 \\ 0 & 0 & 1 \end{matrix}$$

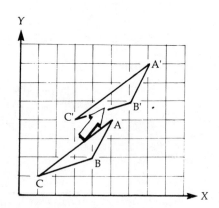

Figure 4.11   *Translation*

The transformation of point A is calculated by

$$
\begin{matrix}
1 & 0 & 2 \\
0 & 1 & 3 \\
0 & 0 & 1
\end{matrix}
\quad \{5 \quad 4 \quad 1\} \quad = \quad \{7 \quad 7 \quad 1\}
$$

Similarly, B and C transform, respectively, to the points

$$\{6 \quad 5 \quad 1\} \quad \text{and} \quad \{3 \quad 4 \quad 1\}$$

The lines bounding the triangle have been translated because of the translation of their endpoints. The result is shown in Figure 4.11.

Rotation

Rotation *about the origin of the coordinates* by an angle τ, where τ is measured positively from the *x*-axis in an anticlockwise direction, may be performed by multiplication by the transformation matrix:

$$
R \quad = \quad
\begin{matrix}
\cos\theta & -\sin\theta & 0 \\
\sin\theta & \cos\theta & 0 \\
0 & 0 & 1
\end{matrix}
$$

Suppose we wanted to rotate the triangle in Figure 4.12 by 30 degrees about the origin, then the transformation matrix would be

$$
R \quad = \quad
\begin{matrix}
0.866 & -0.500 & 0 \\
0.500 & 0.866 & 0 \\
0 & 0 & 1
\end{matrix}
$$

The transformation of point A is calculated by

$$
\begin{matrix}
0.866 & -0.500 & 0 \\
0.500 & 0.866 & 0 \\
0 & 0 & 1
\end{matrix}
\quad \{5 \quad 4 \quad 1\} \quad = \quad \{2.330 \quad 5.964 \quad 1\}
$$

Similarly, points B and C transform, respectively, to

$$\{2.464 \quad 3.732 \quad 1\} \quad \text{and} \quad \{0.366 \quad 1.366 \quad 1\}$$

The triangle has been rotated around the origin by 30 degrees (Figure 4.12).

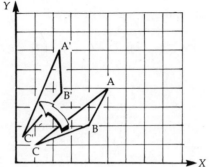

Figure 4.12   *Rotation*

Scaling

Scaling by factors S_x and S_y in the x- and y-directions is carried out by a transformation matrix:

$$S = \begin{array}{ccc} Sx & 0 & 0 \\ 0 & Sy & 0 \\ 0 & 0 & 1 \end{array}$$

Often, S_x and S_y are equal, for instance when we wish to zoom in on part of a drawing. If we wanted to scale the triangle in Figure 4.13 by 2 in both x- and y-directions, then the transformation matrix would be

$$S = \begin{array}{ccc} 2 & 0 & 0 \\ 0 & 2 & 0 \\ 0 & 0 & 1 \end{array}$$

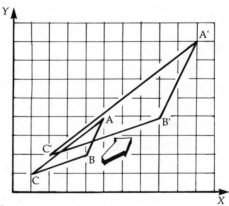

Figure 4.13   *Scaling*

The transformation of point A is calculated by

$$
\begin{array}{ccc}
2 & 0 & 0 \\
0 & 2 & 0 \\
0 & 0 & 1
\end{array}
\quad \{5 \quad 4 \quad 1\} \quad = \quad \{10 \quad 8 \quad 1\}
$$

and B and C transform, respectively, to

$$
\{8 \quad 4 \quad 1\} \quad \text{and} \quad \{2 \quad 2 \quad 1\}
$$

The triangle has been scaled by a factor of 2 (Figure 4.13).

Combining Transformations

Suppose we wish to rotate the triangle in Figure 4.14, not about the origin but about point C. We can do this using three operations:

(a) translate the triangle so that point C is at the origin;
(b) rotate the triangle about the origin by 30 degrees;
(c) translate the triangle so that C is in its original place.

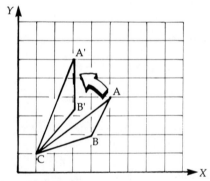

Figure 4.14   *Combined transformations*

The complete operation on point A is given by successive multiplications:

$$
\begin{array}{ccc}
1 & 0 & 1 \\
0 & 1 & 1 \\
0 & 0 & 1
\end{array}
\begin{array}{ccc}
0.866 & -0.5 & 0 \\
0.5 & 0.866 & 0 \\
0 & 0 & 1
\end{array}
\begin{array}{ccc}
1 & 0 & -1 \\
0 & 1 & -1 \\
0 & 0 & 1
\end{array}
\{5 \quad 4 \quad 1\}
$$

The matrices can be multiplied together to give a combined transformation matrix:

$$
\begin{array}{ccc}
0.866 & -0.5 & 0.634 \\
0.5 & 0.866 & -0.366 \\
0 & 0 & 1
\end{array}
$$

Points A, B and C transform to

$$\{2.964 \quad 5.598 \quad 1\}, \quad \{3.098 \quad 3.366 \quad 1\} \quad \text{and} \quad \{1 \quad 1 \quad 1\}.$$

Note that the position of point C is unchanged by the transformations, confirming that it is the centre of the rotation (Figure 4.14).

Three-dimensional Transformations

Three-dimensional transformations are a natural extension of the two-dimensional matrices shown in the preceding section. Again, points in three dimensions are defined in homogeneous form: $\{x\,y\,z\,1\}$.

Translation by T_x, T_y and T_z

$$T = \begin{bmatrix} 1 & 0 & 0 & T_x \\ 0 & 1 & 0 & T_y \\ 0 & 0 & 1 & T_z \\ 0 & 0 & 0 & 1 \end{bmatrix}$$

Scaling by S_x, S_y and S_z

$$S = \begin{bmatrix} S_x & 0 & 0 & 0 \\ 0 & S_y & 0 & 0 \\ 0 & 0 & S_z & 0 \\ 0 & 0 & 0 & 1 \end{bmatrix}$$

Rotations can be performed about the three axes and so three rotation matrices are possible. In the matrices shown, $c = \cos\theta$, $s = \sin\theta$.

Rotation by angle θ about the x-axis

$$R_x = \begin{bmatrix} 1 & 0 & 0 & 0 \\ 0 & c & -s & 0 \\ 0 & s & c & 0 \\ 0 & 0 & 0 & 1 \end{bmatrix}$$

Rotation by angle θ about the y-axis

$$R_y = \begin{bmatrix} c & 0 & s & 0 \\ 0 & 1 & 0 & 0 \\ -s & 0 & c & 0 \\ 0 & 0 & 0 & 1 \end{bmatrix}$$

Rotation by angle θ about the z-axis

$$R_z \quad = \quad \begin{matrix} c & -s & 0 & 0 \\ s & c & 0 & 0 \\ 0 & 0 & 1 & 0 \\ 0 & 0 & 0 & 1 \end{matrix}$$

Projections

Coordinate Systems

Renaissance artists, in order to draw in perspective, would use a device such as that shown in Figure 4.15. In order to display a computer model on the screen, virtually the same method is used (Figure 4.16). The surface of the display replaces the artist's glass plate, the position of the eye is fixed by a set of coordinates rather than by the rather dangerous-looking pointer and often an engineering component replaces the well-nourished model. Much of the calculation involved in producing the image on the screen can be done by using simple three-dimensional transformations. The main complication is in the transformation from one set of coordinates to another. We shall use three separate coordinate

Figure 4.15   *Perspective projection at the time of the Renaissance*

Figure 4.16   *Perspective projection of a computer model*

systems: a three-dimensional right-handed set, a three-dimensional left-handed set and a two-dimensional set of Cartesian coordinates.

World Coordinates.   The component is defined in a right-handed set of three-dimensional coordinates which are chosen by the user. For example, the points on the vertices around the model shown in Figure 4.17 are

A	(1	−1	0)		B	(1	1	0)
C	(−1	1	0)		D	(−1	−1	0)
E	(1	−1	2)		F	(1	1	2)
G	(−1	1	2)		H	(−1	−1	2)
I	(0	0	4)					

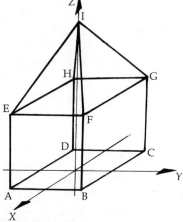

Figure 4.17   *Model – world coordinates*

These points are placed in the coordinate system shown in Figure 4.17. We shall label the three axes W_x, W_y and W_z. The points will be transformed by three-dimensional matrices and so will be put into homogeneous form for the purpose of matrix multiplication; this involves simply adding the fourth coordinate of 1 to each. For example, A will be $\{1 \quad -1 \quad 0 \quad 1\}$. We have not described the model fully, since the lines joining particular points have not been defined but, for simplification, this will be omitted and just the vertex points will be considered.

Viewing coordinates.   The points on the model must be transformed so that they are in a new set of coordinates E_x, E_y, E_z with origin at the viewpoint, E_z pointing towards some desired position on the model and E_x and E_y in the plane of the screen. Viewing coordinates are defined as a left-handed set because their origin is the user's eye and it is natural to assume that distances to points of interest on the model are measured away from the eye. The artist in Figure 4.15 is also using a left-handed set although he is probably unaware of this important fact. To illustrate the method, the viewpoint will be taken at the point (5 6 12) - Figure 4.18. The viewing coordinates of the points on the model must be obtained by a series of transformations.

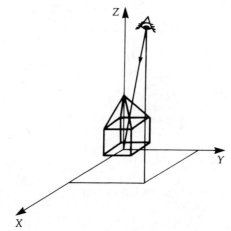

Figure 4.18   *Viewpoint*

Screen coordinates.   These are in the plane of the screen. They are a set of two-dimensional coordinates with origin at the centre of the screen. S_x is horizontal and S_y is vertical. A particular system may use a variety of screens which may be of different resolutions and sizes. It is usual in cases like this to use 'normalized' coordinates, the bounds of the screen being taken as some constant, convenient value which is used for calculation. The true plotting values are found afterwards by scaling the normalized values to suit the particular screen being used.

Perspective

We can use transformations to convert from world coordinates to viewing coordinates. The matrices used are similar to those given in the preceding section except that, since those were for transforming points in a fixed coordinate system and the ones needed are for transforming coordinate systems with points remaining fixed, the signs of the constants must be adjusted.

1. The origin is moved to the viewpoint (Figure 4.19). The coordinate system must be translated, and the transformation matrix is

$$\begin{matrix} 1 & 0 & 0 & -5 \\ 0 & 1 & 0 & -6 \\ 0 & 0 & 1 & -12 \\ 0 & 0 & 0 & 1 \end{matrix}$$

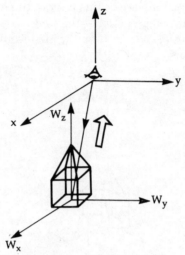

Figure 4.19   *Translate coordinates: new origin at eye*

2. The coordinates resulting from (1) are rotated so that the new x–z plane is parallel with the world coordinate W_x – W_y plane (Figure 4.20). The rotation is 90 degrees about the x-axis and the transformation matrix is

$$
\begin{array}{cccc}
1 & 0 & 0 & 0 \\
0 & 0 & 1 & 0 \\
0 & -1 & 0 & 0 \\
0 & 0 & 0 & 1
\end{array}
$$

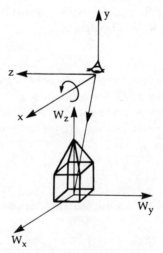

Figure 4.20   *Rotate coordinates: z parallel with W_y*

3. The coordinates resulting from (2) are rotated so that the new z-axis points directly away from the world coordinate W_z axis (Figure 4.21). The rotation is about the y-axis and the angle of rotation is arctan (5/6) + 180 degrees = 220 degrees. The transformation matrix is:

$$
\begin{array}{cccc}
-0.768 & 0 & -0.640 & 0 \\
0 & 1 & 0 & 0 \\
0.640 & 0 & -0.768 & 0 \\
0 & 0 & 0 & 1
\end{array}
$$

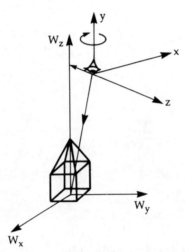

Figure 4.21   *Rotate coordinates: z points away from world origin*

4. The coordinates resulting from (3) are rotated so that the new z-axis points directly away from the world coordinate origin (Figure 4.22). The rotation is about the x-axis and the angle of rotation is arctan (12/61) = 57 degrees. The transformation matrix is:

$$
\begin{array}{cccc}
1 & 0 & 0 & 0 \\
0 & 0.545 & -0.838 & 0 \\
0 & 0.838 & 0.545 & 0 \\
0 & 0 & 0 & 1
\end{array}
$$

5. The new coordinates are changed to a left-handed system - Figure 4.23. This is done by using a scaling matrix with a negative scaling constant. The transformation matrix is:

$$
\begin{array}{cccc}
1 & 0 & 0 & 0 \\
0 & 1 & 0 & 0 \\
0 & 0 & -1 & 0 \\
0 & 0 & 0 & 1
\end{array}
$$

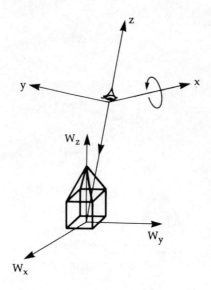

Figure 4.22   *Rotate coordinates: z points to W_z*

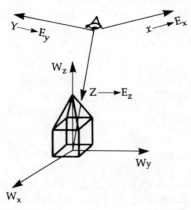

Figure 4.23   *Change hand of coordinates: z points towards world origin*

The body is now in the desired set of viewing coordinates. The transformation matrices of each of the operations (1) – (5) can be multiplied, giving the combined transformation:

$$
T = \begin{array}{cccc}
-0.768 & 0.640 & 0 & 0 \\
-0.537 & -0.644 & 0.545 & 0 \\
-0.349 & -0.419 & -0.838 & 14.318 \\
0 & 0 & 0 & 1
\end{array}
$$

Each of the coordinate vectors of the points A–I of the body is multiplied by this matrix to get the coordinates in the viewing coordinate system (Table 4.1).

Table 4.1   *Coordinates in the viewing coordinate system*

Point	World coordinates			Viewing coordinates		
	W_x	W_y	W_x	E_x	E_y	E_z
A	1	−1	0	−1.408	0.107	14.388
B	1	1	0	−0.128	−1.181	13.550
C	−1	1	0	1.408	−0.107	14.248
D	−1	−1	0	0.128	1.181	15.086
E	1	−1	2	−1.408	1.197	12.712
F	1	1	2	−0.128	−0.091	11.874
G	−1	1	2	1.408	0.983	12.572
H	−1	−1	2	0.128	2.271	13.410
I	0	0	4	0	2.180	10.966

We can now use the artist's procedure to obtain the screen coordinates. The resulting coordinates can then be scaled to an appropriate value. For the moment, we shall assume that from viewpoint to the centre of the screen is unit distance. From similar triangles in Figure 4.24

$$S_x = E_x/E_z$$
$$S_y = E_y/E_z$$

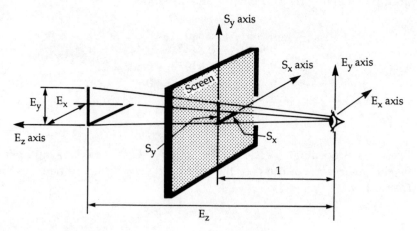

Figure 4.24   *Perspective transformation*

We shall select a scaling factor which will ensure that the perspective image will almost fill the screen; on examining the values of S_x and S_y, it seems that a scaling factor of 1000 is about right for a screen of resolution 512×512. The scaled screen coordinates are given in Table 4.2.

Table 4.2   *Scaled screen coordinates*

Point	E_x/E_z	E_y/E_z	$1000E_x/E_z$	$1000E_y/E_z$
A	−0.0979	0.0074	−98	7
B	−0.0095	−0.0872	−10	−87
C	0.0988	−0.0075	99	−8
D	0.0085	0.0783	9	78
E	−0.1108	0.0942	−111	94
F	−0.0108	−0.0077	−11	−8
G	0.1120	0.0782	112	78
H	0.0095	0.1694	10	169
I	0	0.1988	0	199

As long as the eye position is maintained, the screen can be placed at any distance from it and the perspective image will only change in size. In this case, we have assumed that the image should about fill the screen and should be placed centrally. A typical graphics screen is approximately 360 mm × 280 mm and can be viewed comfortably from about 300 mm.

In general, the picture may need to be clipped before being displayed; in this case, we have been lucky and this is not necessary. Figure 4.25 shows the result. Like all perspective views, this gives the correct impression only when the observer's eye coincides with the theoretical viewpoint.

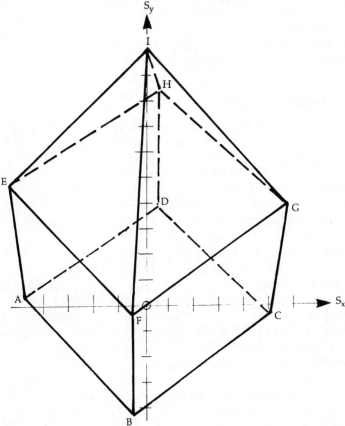

Figure 4.25   *Perspective view of model*

Other Projections

The view just obtained is in three-point perspective which is the most realistic representation of a three-dimensional body on a flat surface. Although a perspective view is realistic and easily interpreted, other projections are often preferred. The conventional engineering drawing is an example. This is sometimes called a multiview orthographic projection: although it takes a fair degree of practice to visualize the physical shape of a complex part shown this way, it is a representation which has served engineers well for many years.

A popular alternative is the isometric projection. This is better for some purposes than perspective, where the foreshortening effect results in lack of clarity of detail. Also, a perspective drawing cannot be scaled easily from the drawing. Isometric projection is an *axonometric* projection; there are others. Pictures drawn in any axonometric projection often seem distorted because distant parts are shown at the same scale as close ones.

Architectural drawings are sometimes in *cavalier* and *cabinet* projections which are examples of *oblique* projections. All these can be processed in much the same way as the perspective transformation by using matrices.

Achieving Realism

The use of perspective and similar projections considerably improves intelligibility when a geometrical model is displayed, but wire-frame models, even in perspective, are sometimes difficult to visualize. It is possible to produce a stereoscopic picture, which is a pair of perspective drawings viewed from two points an eye-width apart that may be examined through a commercial stereoscope. Although this helps to resolve ambiguity and to make sense of a complex model, it is inconvenient.

A more usual method of assisting visualization is to remove the parts of the wire-frame model that would be invisible if the body were solid. In practice, this means that facets, or parts of facets, which are shielded from the viewer by other facets should not be displayed. To remove hidden detail from a body of even moderate complexity needs a good deal of computing and is very time-consuming; it is not to be done lightly.

Hidden line and surface removal is a classic problem in computer graphics and there are many methods documented. It would be inappropriate to discuss any of these in detail here because the ordinary CAD system user normally never has to program a hidden surface algorithm. However, it is instructive to see what are the problems involved.

A body can be described as a set of facets in a world coordinate system which are projected on to a plane surface. If we sort the facets into an order depending on their minimum distance from the viewpoint, then, if we are lucky, they may not overlap and we will have no difficulty in drawing the model. All that we need to do is to draw the nearest facet, clipped if necessary to the viewing window, then the next nearest, clipped if necessary to the viewing window and the facet already drawn and so on.

If we were shading the model (and we shall discuss shading shortly), the procedure would be even easier. We could draw the farthest facet shaded and then superimpose the next nearer one on top of it and so on, clipping to the viewing window where necessary. This procedure is similar to the way an oil-painting is done, and forms the basis of a generalized method called the 'painter's algorithm'.

Of course, when modelling a real-life product, the facets do not behave in such an orderly way and overlaps occur. If we consider the relationship between two facets, several cases are possible (Figure 4.26). In case (a) only the larger facet need be drawn, in (b) and (c) both facets can be drawn directly, but cases (d) and (e) are not so clearcut and need processing.

The aim in hidden line and surface removal algorithms is to identify rapidly those cases which do not present complication; this is much the same as the reasoning behind the clipping algorithm which was discussed previously. Most of a picture to be displayed requires little processing, it is usually only a small proportion that requires complex processing and if we can identify these parts rapidly then the result is a nett increase in display processing. Another good principle, that of 'divide and conquer', is also used in many of the algorithms. If it is impossible to process a view easily, then the viewing window is subdivided repeatedly until a simple case occurs, if necessary going down to the pixel level.

It is interesting to note that a human being faced with the problem of removing hidden lines from a wire-frame picture would be as lost as the computer and could not do it without prior knowledge of the solid form. From our knowledge of the real world, we might guess that Figure 4.27a would look like Figure 4.27b in the solid, since regular shapes occur more often in engineering than irregular ones, but Figure 4.27c is also a possibility. In fact, there is an infinite number of shapes that would have a projected view like Figure 4.27a. Since the computer model is held unambiguously, this is not a problem in automatic hidden line removal. The only difficulty is one of achieving sufficient speed.

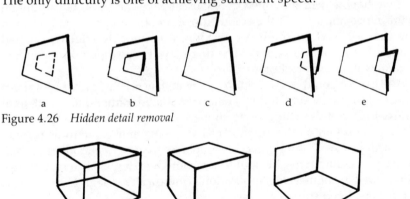

a   b   c   d   e

Figure 4.26   *Hidden detail removal*

a   b   c

Figure 4.27   *Ambiguous perspective views*

Shading is also a useful process for increasing the viewer's information about the shape of a body. If we know the details of the facets, or portions of facets, that are visible to the viewer, then sufficient data exists to provide a shaded model. In order to shade a body at all realistically, it is necessary to consider all the ways in which light is reflected from it to the eye of the observer. Firstly, in most real environments, the body is lit by *ambient* light, the light which it receives from its surroundings. This is the same at all points of the body. Secondly, it is usual to have at least one light source from which the body receives light and reflects it to the viewer's eye. This is *diffuse* light. Diffuse light may be thought of as being reflected from beneath the surface of the body so that it takes up the colour of the body and its intensity depends on the angle at which it falls on to a facet and the angle that the facet makes with the viewing direction.

Both of these effects must be taken into account if the shading is to be at all realistic. If only ambient light were modelled, the body would have the same shading at all points; a sphere would look like a disk. If only diffuse light were modelled, any part of the body not receiving light directly from the light source would appear black. If just these lighting effects were considered, the model would look reasonably realistic but perfectly matt.

If it were not intended to model a matt surface, but a metallic or plastic one, the lighting algorithm would also have to take account of *specular* reflection ('specular' means 'mirrorlike'). This is light reflected from the surface of the body and is of the colour of the light source. This is rather more complicated than the other two effects since its form and intensity depend on the shininess of the body being modelled. A shiny body will have intense highlights of small area while a matt body will have larger but dimmer highlights. There are several models for specular reflection; one still widely used is one of the first devised: Phong's model.

If we have a model defined geometrically we can thus, by supplying extra information about the colour, position of the viewpoint and light sources, and surface shininess, produce a reasonably effective shaded model. The smoothness of the shading may also be enhanced by one of several well-known techniques.

And so we could go on, adding, for instance, shadows and surface textures; varying visual depth of field; distributed, rather than point, light sources. Currently, there is no unified technique for dealing with all visual effects (although a method called 'ray-tracing' can handle most of them) and each commonly receives separate treatment and takes considerable processing time. Most of them, though extremely interesting, are not needed in engineering applications where a lesser degree of realism is often considered sufficient.

Problems

1. The version of Bresenham's algorithm described applies only to lines with slope from 0° to 45°. Verify that this is true, and modify the algorithm so that it takes all other cases into account.

2. Find transformation matrices which will perform the following actions

on the triangle shown in Figure 4.11:

(a)   Rotate it 90° about its centroid.
(b)   Mirror it about the X-axis.
(c)   Mirror it about the X-axis, then about the Y-axis.
(d)   Mirror it about the line $Y = X$.

3. Write a small BASIC program which will take as input the coordinates of two endpoints of a curve and the coordinates of two control points and display the resulting Bézier curve. Make the program interactive, so that the positions of the control points may be varied to give different forms of the curve.

4. The following terms are used in graphics. Write a short definition of each in your CAD dictionary.

(a)   Fractals
(b)   Ray tracing
(c)   Gouraud shading
(d)   Mach bands
(e)   Depth cues
(f)   Half-toning

5. Another elementary transformation that we did not describe in this chapter is that of shearing. A shear matrix looks like this:

$$\begin{matrix} 1 & a & 0 \\ b & 1 & 0 \\ 0 & 0 & 1 \end{matrix}$$

Investigate the action of this matrix by trying various values of a and b, and transforming a unit square with one corner at the origin of the x–y coordinate system.

5   Management issues

Introduction

'Counting the Cost'

Scenario

We are back at Coolpoint Ltd a few weeks after our last visit. The people that we met before are attending another meeting and have been joined by the Chief Accountant. They are about to discuss again whether they should invest in a computer-aided draughting system. This time, they have a little more information and have engaged a consultant to help them with the decision. He is also present:

MD: At the last meeting, we were left a little up in the air as to our future action. Since then we have gathered more data and I think we should soon be able to come to some sort of a decision. The Chief Designer and his people have looked at the systems available and there seem to be several systems which will suit our modest needs. He seems to have homed in on MODCAD. Perhaps he could kick us off. Briefly.

CD: We have decided that, in the first instance, we need six workstations running a draughting system. Of course, when we become proficient in its use, we might then consider the purchase of a three-dimensional modelling system and other goodies but the priority is draughting. There are three systems possible; I incline towards MODCAD for technical reasons which I can detail if it is your wish. Roughly, the cost would be £100 000 for hardware and £50 000 for software. I should like to stress once more that Watussi, our main competitors, have spent nearly a million pounds on their system.

MD: That doesn't seem too bad to me. I'm sure the Chief Accountant would like to ask you a few questions. [*Jocularly*] Right. Your witness.

CA: You say that we should need to spend £150 000 on the system?

CD: Yes, six workstations should just about fit into our cost envelope.

CA: I think that we should look at it in a little more detail. What do we do if the system breaks down? I believe they have been known to. Do you all go home?

CD: Yes, you are quite right. We could do with a guaranteed 24-hour maintenance contract. This would be an extra 10% of hardware cost per annum. We do get free maintenance for the first three months, though.

CA: I'm glad to hear it. And would the vendor supply us with any updated versions of the software? I hear that these are not uncommon.

CD: A software maintenance contract would be 12.5% per annum. They would also provide us with a telephone query service as well as updating our software.

CA: I put it to you that this, if my calculations are not in error, would involve us paying £7500 the first year and £10 000 in subsequent years for hardware maintenance, and £6250 each year for software maintenance. Is this correct?

CD: [*Mumbling*] Yes.

MD: Would the defendant please speak up?

CD: *Yes!!*

CA: Now you are on record as stating that you need six workstations for our twenty DO personnel. Does this imply that your staff only spend about one third of their time drawing. Five hours a day so-called 'thinking time' seems a bit excessive. I doubt if I spend as long myself.

CD: I object to the slur that has been cast on the diligence of my staff.

MD: It seemed a reasonable comment to me. Can we push on?

CD: To be quite honest, we are considering running a two-shift system like Watussi have done. They claim that they have …

CA: But haven't they given their staff a 20% pay increase to compensate for the anti-social hours they will be working every other week?

CD: I suppose that we would have to do the same.

CA: Average pay in the DO is about £12 000. The wage-bill would increase by £48 000 each year. We would also need to nominate a supervisor for the extra shift. That would be an extra £2 000 per annum at least. Now let us turn our attention to the equipment. You have stated, and stated on oath, that the system will 'almost run itself'. I should like to examine your interpretation of the word 'almost'. Would we not have to make regular and frequent security copies (I believe the technical term is 'back-ups')? And I suspect that there are other activities that your staff, even with their talent and adaptability, could not be expected to perform.

CD: Yes, back-ups are essential.

CA: So then we would need someone responsible for doing these tasks on each shift.

CD: No, not necessarily. We were thinking of just using one person on the late shift, for a while at least.

CA: I wish I were as optimistic as you seem to be. Well, let us assume that what you say is correct. It would entail employing an extra member of staff. With employer's contributions, that should cost us £20 000 a year. Now let us consider training. Would you describe, in your own words, how we are to go about training your twenty people so that they can use the system as rapidly as possible?

CD: We are given two free places on a training course as part of the deal. I think that at a pinch we could just send the two section-leaders who could then train their people themselves.

CA: At a pinch, eh? Wouldn't this 'pinch' mean that, even if your two worthies were capable of training the other staff effectively (which I doubt), the time waiting for the system to become productive would be undesirably long?

CD: Yes. It usually takes about three months for a draughtsperson to get up to the normal speed on a system. We *would* be better off if we sent everyone on a course. The cheapest way would be to get the vendors to do the courses here. They charge £300 a day for groups of six maximum. They run elementary and advanced courses, each lasting five days. I would like to send everyone on the elementary course and just the two section-leaders on the advanced course after about six months.

CA: Another £6000, eh? Not to mention the time lost in drawing production. Now you have stated that the system will improve our corporate image. I assume that you are going to show it off to customers. Are you going to put it in a corner of the Drawing Office?

CD: No, we could have part of the DO converted to a CAD suite. We need it partitioned off, we need the lighting changed and some good special-purpose office furniture and...

CA: I think I get the idea. Let's say £20 000 will cover it. Running costs: stationery, magnetic media and so on would cost, you have stated, £5000 a year. Now I could continue mentioning rates, electricity and other costs but I might run the risk of being called obstructive. I have been doing a few sums and, with the MD's permission, I should like to break off for a few minutes to get photocopies for the jury... [*flourishes the notes shown in Figure 5.1*].

                    ■   ■   ■

COOLPOINT
Laundrymaid to the World

Rough cost analysis based on 3 years.

Hardware Cost	100 000
Software Cost	50 000
Interest	50 000
Hardware maintenance	27 500
Software maintenance	18 750
Shift bonus	144 000
Supervisor	6 000
Operator's wage	60 000
Training	6 000
Running costs	15 000
Site conversion etc	20 000
Total over 3 yrs.	497 250

Cost per annum = £165 750

Workstation hours/annum = 2000 × 6 × 2 × ·9
= /21 600

(assuming 10% downtime).

Workstation cost = 165750/21600 = £7·67/hour

Say £8/hour !! since some other costs not included.

Figure 5.1   *Typical 3 year cost analysis for a CAD system*

The figures shown in Figure 5.1 are fictitious but not untypical. If we look at percentages, it will be noted that the hardware and software are a comparatively minor part of the total annual cost, even though the Chief Accountant expects them to pay for themselves after only three years. Three years is not an unreasonable period as a basis for this kind of calculation; the time allowed usually lies between two and five years. It is also common to add to the expenditure the interest which would have been gained if the money had been invested. In this case, the Chief Accountant has been kind; it seems that he has worked on a basis of only 10% per annum.

The aim, as with any plant, is to achieve 100% utilization, which is the reason why many installations are operated on a shift basis. Workers on production commonly work three shifts but since shift-working is a fairly new phenomenon for white-collar workers, it is usual to have two shifts. Another possibility is for staff to work longer hours but fewer days. For instance, staff are divided into two groups and each group works three days, twelve hours per day. This has been tried in France and there are indications that it is more effective and more popular with staff than a two-shift system.

It would appear from the analysis that it would cost comparatively little to add an extra workstation or so to the installation. A problem here is that cost often does not increase linearly. One pays a lot for the first workstation, then unit costs decrease and extra station costs are cheap. However, after a certain critical number there is a sharp increase in the processing power needed. In the same way, extra support staff are needed to cope with the added equipment. Presumably, the Chief Designer has taken these factors into account.

'The Empire-builder Strikes Back'

Scenario:

Back at Coolpoint again, half an hour after our last visit – the Chief Designer has had time to recover his composure and is deep in discussion with the consultant:

MD:	I think that we have all looked at the figures. Perhaps the Chief Designer would like to comment.
CD:	My learned friend has pointed out that the system will cost rather more than he anticipated. I should like to explore the benefits that we would obtain from the system. With your permission, I should like to call an expert witness - Mr A. Consultant.
MD:	Take the stand, take BS308 Part 1 in your right hand and repeat after me ...
CD:	Mr Consultant, you recently conducted a survey of the activity in our Drawing Office, did you not?
AC:	Yes, I took a sample of the drawings produced. These were as follows:

Electrical schematics	20%
Assembly drawings	10%
Detail drawings	35%
Sheet metal drawings	20%
Tool designs	15%

CD:	That seems about right to me. If we introduced a draughting system, how would productivity increase in each of these categories?

AC: It is difficult to be accurate but, in my experience, drawing productivity ratios for the types of drawings that you do are:

Electrical schematics	4
Assembly drawings	3.5
Detail drawings	3.5
Sheet metal drawings	3
Tool designs	1.5

This gives an overall productivity ratio of
$(0.2 \times 4 + 0.1 \times 3.5 + 0.35 \times 3.5 + 0.2 \times 3 + 0.15 \times 1.5) = 3.2$
I must stress that this is on the conservative side.

MD: You may cross-examine the witness.

CA: I expect that you are right, Mr Consultant. By the way, I forgot to include your fee in my rough costing. I should like to ask your opinion on one or two points of detail. You claim that we can achieve a productivity ratio of 3.2. Will this be immediate?

AC: No. The time taken for people to attain their potential varies, but most can draw as fast on a computer system as they can manually after three months. I should guess that all the staff should reach the figure I have given you after about eighteen months.

CA: So - and correct me if I am wrong - over our three year period, the actual productivity ratio will not be 3.2, but about 2.4?

AC: Perhaps a little higher. I did say that the figure was conservative.

CA: Thank you. Let us call it 2.5. Now, I don't know much about draughting, but on the too few occasions when I have visited the Drawing Office, it seemed to me that not all the staff were engaged in actual drawing.

CD: May I come in here? We estimate that a third of DO activity is in preparation, consulting other people, getting drawings from the store, doing preliminary calculations and so on. To forestall your next question, it may well be that CAD will reduce this time eventually, but we wouldn't expect it to have much effect over the next three years.

CA: So the productivity ratio of 2.5 only affects two-thirds of the drawing time? I estimate that we are now down to a ratio of about 1.9. Let's be generous and say 2. The cost of each of the six workstations is £8 an hour. We could take on an extra eight staff or employ contractors and increase drawing throughput by 40% with no trouble at all, couldn't we? And if we employed younger staff to do the easier tasks, I should think we could get close to your figure without all the organizational upheaval.

CD: Well, yes, but...

CA: I rest my case.

■   ■   ■

The argument could go back and forth for a long time. The discussion illustrates the difficulty of claiming improvements in crude productivity or drawing throughput as a reason for committing a firm to CAD - or, indeed, most kinds of new technology. The most valid reasons for investing in CAD are not directly financial at all; they are what are sometimes called 'intangible' benefits, although in this context 'intangible' is a poorly chosen word. The benefits are tangible enough but are not easily expressed in profit. On the other hand, it *is* useful to perform a cost analysis of the kind outlined by Coolpoint's Chief Accountant, if only to demonstrate that productivity is liable to increase with the new system. It is wise, for several reasons, to treat these analyses with caution. For one thing, they are based on estimates; nobody can be sure beforehand what the increase in drawing throughput will be. There are tables widely quoted showing the probable productivity ratio for different classes of drawing, but these are crude approximations and take no account of the capability of the persons using the system.

The true benefits of CAD are not easily quantifiable. Some of the major ones are the following:

Better drawings.   A draughting system can be configured to a firm's liking and the resulting standard easily imposed on users. Most firms that are seriously committed to draughting systems have standards for such features as dimensioning, hatching, conventional symbols, layer utilization and standard notes. Although there is such a thing as 'drawing style' in computer-aided draughting, the use of enforced DO standards makes drawings more uniform and, if the standards have been chosen well, more easily understandable. Visualizing a complex component from a conventional orthographic drawing is often difficult and needs much practice, but many systems support the creation of perspective and isometric form. Even a simple wire-frame model is an enormous help in interpreting what a component looks like in the round. And since a drawing is the engineer's primary means of communication, any improvement in intelligibility is worthwhile.

Control of information. Successful management depends on being able to obtain high-quality information quickly, so that progress can be monitored. CAD permits this because data can be held centrally and can be accessed rapidly and in a controlled way.

Control of modification. In many (or probably most) DOs a considerable proportion of time is spent in modifying drawings. This is often done by altering prints: if several modifications have been carried out in this way, the end result is often an out-of-scale and unintelligible drawing. It is certain that computer-aided draughting is ideal when drawings are to be modified. Old drawings can very easily be taken from the archive files, modified and re-named. The result is an accurate, clean drawing which is of as high a quality as the original.

Communication. The conventional drawing is usually the most concentrated source of information about a product. On it are, typically, a definition of the product's physical shape, a specification of the material from which it is made, manufacturing and gauging information and, in the case of assembly drawings, a parts list. These are of interest to many other departments in the organization - production planning, production scheduling, inspection, costing, weight control, stress analysis and others.

In the manual system, prints of drawings are circulated and often mistakes are made because people are working to the wrong issue number. If the drawing is placed in a system archive, then issues can be strictly controlled. Also, drawings can often be passed directly to some other system such as computer-aided manufacture. Some parallelism in design is also theoretically possible for urgent projects. For instance, checking of drawings might be carried out while they are in progress and more than one person might work simultaneously on the same project - indeed, on the same drawing, although there are reasons why this might be considered undesirable.

Improved image. Firms sometimes go into newer technology for no better reason than that they think they ought to. In our example, Watussi have made a decision to do all their drawing on a draughting system. Coolpoint feel that they have to follow suit. They regard the CAD system as a symbol of technical modernity, hence of efficiency. It seems likely that this reason, though not often stated outright, is very common. When firms are visited, generally the visitor is taken round the CAD installation which seems to be viewed by management in much the same way as they regard the reception area - the workstation replacing the familiar potted palm. This is not so foolish as it might appear, since the involvement of a firm with CAD is a sign that it *is* forward-looking and innovatory.

Enforced decision.   Large firms, and car manufacturers are an example, use many suppliers. They sometimes insist that suppliers' drawings are provided in a form that can be input to a CAD system. In this case, there is little that the supplier can do but buy a suitable system.

Job enrichment.   Although it is difficult to imagine a firm buying a system purely for this reason, it is true that the majority of users enjoy drawing on a draughting system. One has only to attend a users' group meeting to see how users identify with their system, exchanging handy hints and experiences. It seems probable that some of the high increases in drawing throughput claimed by vendors are due not to the excellence of the system but to the users working harder on them.

These are some of the *true* benefits that CAD brings. The problem is to present them in a plausible case for the purchase of a CAD system.

System Ease of Use

It is appropriate at this stage to discuss the features of a system which will enable a firm to perform the transition from a manual to a computerized system. A critical factor is the speed with which personnel can adapt to the system; this to a large extent depends on the care that the system designer has taken with the user interface. Firms are often disappointed by the slow rate of increase of drawing productivity when a draughting system is installed. To achieve a ratio of one-to-one with the manual system can take three to six months. This 'burn-in' period depends on the type of drawing, on the user's aptitude and, of course, on the quality of the system being used.

There is little that a firm can do, beyond giving training, to improve a user's rate of progress, although some of the larger companies may run a 'closed shop' system, where the CAD installation is not available to all drawing office personnel but is only used by staff who have been aptitude tested. The major influence on the speed with which a user can become proficient on the system is the quality of the 'Human–machine interface'.

In Figure 5.2, on the left is a human being whose most natural means of communication is by word, facial expression and gesture, and on the right is the machine whose most natural means of expression is a stream of ones and zeros. The human likes to speak in a natural language such as English, the machine prefers machine code. This communication gap is bridged by the users communicating not with the bare machine but through the medium of a program such as a three-dimensional modeller. At the moment, it is impossible to move completely to the left and let the user converse with the system in a natural language; some think that it never will be possible. It is necessary for the system designer to compromise and restrict the user to a set of fixed format commands which, in a well-designed system, are as natural as possible. Unfortunately, in many systems this attempt towards naturalness results in verbosity: an interface that is very helpful to a novice can be extremely irritating to a more skilled user who generally prefers commands to be more terse.

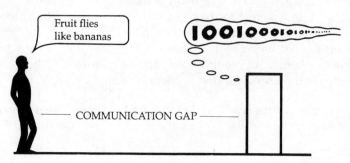

Figure 5.2   *Human–machine communication*

Because the design of the human–machine interface has such a marked effect on speed of training and subsequent productivity, considerable research is being done into the nature of good interfaces. These largely draw on psychological theories of human cognition and it is inappropriate to discuss them here. We can, though, make some general comments about the design of good user interfaces.

Quick response is one of the most critical factors. If users have to wait a long time before the system acts on commands, then their concentration is interrupted and they become impatient. An example of this is when a time-shared system is heavily loaded and the computer does not echo typed characters immediately. The lag between striking a key and the character appearing on the screen is very disconcerting and typing errors usually result.

On a CAD system, the user's expectations vary for different system activities. If the activity seems easy, then the user expects a response very quickly - two seconds is the figure usually quoted. On the other hand, if the operation looks difficult, then the user is more tolerant. The problem is that what seems easy to the user might involve a lot of computing - deleting a line is an example. The converse is also true. For the system to re-draw a picture is comparatively easy, but since users would take a long time to repeat the drawing manually, their expectations are not so high. It is common to pacify the user, if a long task is being performed, by displaying a 'placebo' which indicates that the system has not seized up and that some processing is still going on. A clock with rotating hands is popular.

Adaptability is the ability of a system to provide varying levels of interface to suit the skill of the user. Novice users may be given very detailed prompts explaining clearly what is to be done next. Advanced users are given either terse prompts or none at all on some systems. A typical draughting system permits input by a menu-pick, or by typing commands on the keyboard. Novices tend to use, almost exclusively, the menu for their commands. Advanced users tend to prefer to type commands, or to use a combination of input by the keyboard and menu-picks.

A structured screen layout can affect ease of use, also. Screen displays contain various types of information. On drawing systems, the screen is partitioned into areas, each having a special purpose - for example, in AutoCAD a strip down the right-hand side is reserved for the menu. Across the screen, at the top, is an area which holds the drawing scale and system status. The strip at the bottom is used for displaying commands and dimensions. The rectangle in the middle of the screen is used for drawing. Because these are kept in the same places throughout, the user's eye automatically finds them after a little practice. More and more systems are using colour as a means of identifying different types of information; this is a great help in attaining speed of working.

Naturalness of a system is the closeness with which the system follows the manual activity. A reason why many users find difficulty in adapting to a system is that they have to change their working procedures. Broadly, the less they have to change them, the better for rapid user adaptation. One problem is that CAD systems can do some things that cannot be done manually - on draughting systems, views can be moved around, copied, rotated; symbols can be retrieved; drawings can be multi-layered. It is difficult to make these and similar operations natural.

It must be admitted that, with the exception of draughting systems, many CAD systems have poorly designed user interfaces. If the same degree of attention were given to detail that seems to have been paid to the design of 'Wimps' (windows, icons, mice and pointers) on some micro-based business systems, then the training of CAD personnel might be much quicker.

Selecting the Right System

The success of a new CAD system depends on how closely it matches a firm's needs and it is usual to select the system with care. It is customary, when a substantial amount of money is to be spent, to carry out a feasibility study. A detailed description of the stages involved may be found in books on system analysis; we shall give a brief account of the procedure.

It is usual to set up a task group to investigate the procurement of the system. The team is made up of representatives from all the major departments that have a direct interest in the system. We have already argued that the full benefits of CAD will only be obtained when it is connected to other systems such as CAM, and it is likely that there will be interested parties other than design personnel involved - for example, manufacture and the computing department. It is also useful to have a member with some financial expertise. The most effective size of the group is a matter of compromise between obtaining sufficient representation of all interested departments and keeping it compact enough to be manageable. Since there is a possibility of conflicting interests, it is important to select a suitable chairperson. If the firm is completely new to computer systems, then it may be beneficial to obtain the services of an independent consultant to advise or to head the group.

The first stage of the study should be to gather information about what the system is expected to do - a 'needs analysis'. It is necessary to define the precise functions required and to analyse the expected workload. Since there is broad representation on the task group, this stage can be delegated, but some delicate negotiation between the various interests may be needed. The information gathered might be quite detailed, the future drawing requirements being estimated by sampling Drawing Office activities.

When the needs have been defined and agreed upon, a system specification can be drawn up. The refinement of the specification depends on the amount of detail that has been obtained in the needs analysis, and this in turn depends on the time and effort that have been spent on it. Unfortunately, this sort of exercise is seldom a leisurely affair and the specification is often fairly sketchy. It should, at least, have a functional description of the hardware and software needed. It would be ideal, for instance, to be able to specify, with certainty, that we need 'a pen plotter with four pens, changeable under software control; capable of plotting up to A0 size at a pen speed of 1 m/s...' and so on, but usually the specification is more on the lines of 'capable of producing high quality plots up to A0 size'. It is rare to have even the processor specified in detail beyond the number of users that it must support.

Based on the specification, possible systems can be selected. There are many CAD systems on the market at any one time, but because of the limitations of cost and the field of application, the task is not so daunting as it might at first appear. There are several directories of software widely available - the CADCAM Association's *Yearbook* is an example - and specialized journals are also a good source of information. It is also possible to find out which systems are used by other firms in similar fields. It may be that no available system exactly matches the specification; in this case, it is useful if the detailed hardware and software functional specifications have been prioritized or some weighting method has been used. It is also common for more than one system to satisfy the specification. In any case, the vendors of candidate systems are invited to tender.

On receipt of the quotations, a fairly detailed costing may be carried out. This should be the total cost - a rather more precise version of the one done by Coolpoint's Chief Accountant. This should include not only one-off or 'non-recurring' costs such as the hardware, but also recurring costs such as the hardware maintenance. The candidate systems can then be ranked in order of merit and, if they all satisfy the specification, then the cheapest is the 'best'.

It is usual to be cautious and check the choice of system by performing 'benchmark' tests. These are jobs which are chosen as representative of the work to be done on the system in practice and which are run on the system, often at the vendor's premises. The aim of benchmarking is to confirm the vendor's claims about the system's performance; if it does not perform satisfactorily, then the next 'best' on the list is bench-marked. If no system satisfies the firm, then the whole process of selection may be repeated. Possibly the specification is too rigorous for the budgeted figure.

Installing and Running a System

The Equipment

At last the decision has been made. A system has been chosen and the order has been placed. The events which follow are:

The site is completed to everyone's satisfaction
The CAD system is delivered
The system is installed by the vendor
Acceptance tests are carried out
Work commences

The vendor will usually give advice on the site preparation. Some of the factors to be considered are layout of workstations, modification of lighting and installation of lines and power supplies. Although the factors contributing to a good working environment are well documented, it is rare to find an installation which is ideal. The most common shortcoming is in the lighting, which may be inadequate on working surfaces and placed incorrectly so that glare is reflected from the screen. Faults such as these are often the result of expansion of the system beyond the planned level and the subsequent over-crowding of the working area.

Even in the smallest of installations, where a solitary stand-alone system is placed in a corner of the drawing office, it is important in the interests of drawing productivity and user comfort to make some effort to create a good working environment. In a medium-to-large installation, it might be considered advisable to separate the main computer from the user area; the plotter might also be remote from the workstations. The planning of a room for a large computer is a specialized business: false floors may need to be installed, air-conditioning and temperature control may be required, special power supplies needed and so on. Informed advice is a necessity for all but the most experienced of firms. The environmental requirements for smaller dedicated turnkey systems or networks are fortunately not so stringent.

When the site has been prepared, the system is delivered, unboxed and installed by the vendor. It is thoroughly tested before being accepted by the buyer. Then, of course, the real problems start because people are involved. Some of these problems will be discussed.

The Staff

Setting Up Training Programs

It is useful to have general awareness training sessions for all the staff involved, even if not directly, with the system. To get the maximum benefit from computers, it is necessary to have (or, at least, to aim for) a company-wide integrated system and a little training at the initial stages of a firm's involvement with CAD will pay dividends later in the system's development. The major training requirement is, however, that of the potential users. Since it is highly desirable to start production imme- diately the system has been accepted, training should have started some time previously (although not too long before).

CAD Staff

In an installation of any size, it is important to have someone responsible for the day-to-day organization and running of the system. In a sizeable installation, this may involve the appointment of a CAD manager. It is a source of some debate whether it is better to train existing staff in the ways of computers or to appoint a computer expert from outside to be trained in the ways of the company; the consensus of opinion seems to be that the first course is better, if a suitable person can be found. Ideally, the manager should have some involvement in the planning of the installa- tion and the setting up of working procedures and so should be appointed before the computer goes live.

The size of the computer staff will depend on whether the system is centralized or distributed, on the amount and complexity of equipment and whether a shift system is in operation. Vendors often claim that their systems just need the power to be turned on and drawing to be commenced but, in practice, this is seldom true except in the most primitive of systems. There always exist such jobs as loading of plotters with pens and paper, and hard-copy units with toner, mounting magnetic tapes, keeping system records and fault reporting. It is counter-productive for design staff to spend time doing these, except in a very small installation. In larger set-ups, there may be several operators under a chief operator and shift-leaders.

If a company is lucky, the system as bought exactly matches its needs. As we have previously discussed, a successful CAD system is not just a collection of independent programs but should have intercommunication between its component parts and also with other parallel systems such as CAM. There are large and comprehensive integrated systems on the market, but none of these might be considered suitable or, if one is bought, it might be necessary to tailor it to the company's special needs. In either case, subsystems must be patched together to make a working system and skilled support is needed to perform this 'system-building' task. This requires a fair knowledge of systems and application programming. Again, if a firm operates in a very specialized field, there may be no analysis programs available 'off the shelf' so that it is necessary to develop them in-house. In either case, programming support is needed. The amount of this support will determine whether consultants or full-time staff are required.

The installation will eventually have a drawing archive from which past and current drawings may be retrieved for reference and modification. Most firms have a lot of drawings stored away; when a CAD system has been implemented, a decision must be made whether the manual system of holding copies on paper is to be continued, whether drawings are to be stored on magnetic media such as tape, or whether a combination of both methods is to be used. If they are to be stored on tape, then they are normally held in a tape library with some sort of procedure, possibly computer-based, for indexing drawings and controlling receipts and issues. The number of drawings may require a tape librarian to be employed.

CAD can improve the quality and uniformity of the drawings (and also the products) produced, but to obtain this benefit it is necessary to consider drawing standards. Many firms, in the interests of efficient communication and reduction of misunderstanding, impose a drawing standard such as BS308. A good deal of standardization can be achieved through the draughting system, by providing standard tolerance-boxes and machining symbols, by having a company standard for layer utilization, by having set pen-widths for different purposes and so on. These standards must be set up and documented.

Often, the supplied system documentation still falls far short of the ideal. Improvements in standards of documentation have been noticeable over the last few years, but it is still common to find unintelligible manuals supplied with systems and it may be considered advantageous for the company to prepare its own low-level documentation specific to its needs. For most large systems, a good deal of documentation is provided at different levels. There is usually a 'Getting started' manual; a more advanced and comprehensive manual; a definitive list of system commands and their effects; a configuration manual; a systems programmer's manual. These are subject to frequent modification as the system develops and new versions are issued. Many large firms employ a documentation clerk to ensure that the information held is of good quality and is available when needed.

If the CAD system is part of an integrated system, it will be used to set up a good deal of the database which is the hub of all the other systems. The success of the system depends crucially on the quality of the database and so it is vital that it should be maintained in an efficient way. Not only will a database contain details of designed products, there will also be subsystems for selecting such features as materials and machine components. The importance of databases has been discussed; firms highly committed to integrated systems have staff specially employed for working on them.

Of course, comparatively few CAD installations will need all the staff that have been detailed here, but it is certainly true that some of the activities described will be necessary in even the smallest firm. They should be considered in the planning of the system organization.

Problems

1. 'Benchmarking is useless because it tests the operator rather than the system. A poor system used by a good operator may show up better than a good system used by a poor operator.'
Discuss the truth of these statements.

2. Perform a complete costing on the system that you use based on the factors mentioned in the scenarios 'Counting the cost' and 'The empire builder strikes back'.

3. A report issued by professional associations in West Germany, France and the UK has stated that a major obstacle to the adoption of microprocessors in British industry is the reluctance of middle management to accept them. The move towards integrated sytems is likely to be accompanied by a decrease in departmental specialization and responsibility. It is possible that middle management will be even more reluctant to accept integrated systems than they have been to accept the use of microprocessors.

Write a short scenario in which a Drawing Office manager fights a rearguard action against the incorporation of an integrated system.

4. Assess the 'user friendliness' of the system that you use, based on the criteria mentioned under 'System ease of use' in this chapter might be better.

5. Write a case for the Chief Designer of Coolpoint Ltd, based on the benefits that a CAD system would bring which cannot be costed directly.

6 CAD standards

Drawing Transfer

'Pressing Matters'

Scenario

Coolpoint Ltd make domestic washing machines. They buy in most of the components for their machines which they assemble. The main supplier of their decorative trim is Palaeolithic, a small firm which specializes in pressings of this kind.

Coolpoint have had a large and expensive CAD system for several years and their component records are all maintained by the system. Palaeolithic have a different CAD system, smaller than Coolpoint's and used mainly for draughting.

Palaeolithic supply drawings of their pressings to Coolpoint on paper and they are entered into the CAD system by re-drawing them. Not only does this waste a lot of time, it also leads to a lot of copying errors and Coolpoint are not at all happy with the procedure. They now insist that drawings are supplied, not on paper, but in a form that can be input directly to their CAD system.

This scene takes place in the office of Mr A. Stickler, Chief Engineer of Palaeolithic. Mrs D. Wheal, the Chief Designer, enters:

AS:	Good Morning, Daisy. Have you heard the latest from Coolpoint? They want us to send our drawings on magnetic tape.
DW:	Yes, I've been expecting it for some time now. I believe it's quite a common practice these days and anyway, you can see their point.
AS:	Let's humour them, then. We normally store our drawings on disks, don't we? Doesn't sound much of a problem.
DW:	[*Tactfully*] Well, I suppose we could, but I don't think it would be much use to them. We use GrotCAD which stores drawings in a completely different way from their system. So even if we did send them a disk with a drawing on it, their system wouldn't be able to make much sense of it.

AS: Why didn't we buy the same system as them in the first place? I don't like reminding you, but I did say at the time that we were rushing things when we decided on GrotCAD.

DW: I would have jumped at the chance of having the same system as theirs, but I was given a ceiling of £50 000, if you remember. They have invested three-quarters of a million in their PlutoCAD. In any case, what about Watussi? They are as big a customer as Coolpoint and they have an even bigger Interdebt system.

AS: Isn't that typical of computers! Why can't they all be standard?

DW: You might as well ask why washing machines aren't standard.

AS: Anyway, what can we do about it? Couldn't we get someone to write a program to translate from one drawing to another? Our programmer doesn't seem to be doing much lately and he made a good job of that mailing list program last month.

DW: It's not as easy as it sounds. The way computers hold drawings is a bit more complicated than a name and address file. It would take a matter of months, I guess.

AS: [*Gloomily*] And I suppose we would finish up having to write translator programs for all our major customers. With our luck, they would all be different.

DW: They are, as a matter of fact. *And* we would probably have to update them each time there was a revision in one of the systems.

AS: We didn't have this trouble with the drawing-boards and tee-squares. Well, come on, is there any other way? I am sure that you are dying to tell me.

DW: Luckily there is. GrotCAD supply a program which translates from their format into a standard format called IGES - the Initial Graphics Exchange Specification. It doesn't cost much - well, a lot less than it would cost to write our own translator. And I know that all the large systems, including those used by our major customers, have translator programs from IGES to their own formats.

AS: ? [*Looks puzzled.*]

DW: Let me show you. Suppose we wrote our own translator
             programs for each of our four main customers. We would
             have to write four programs like this: [*sketches Figure 6.1*].

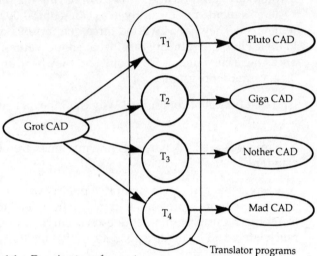

Figure 6.1   *Drawing transfer – out*

DW: Also, although you haven't mentioned it, Coolpoint want
             to supply *us* with their drawings on tape. So we would
             need another four translator programs *to* GrotCAD: [*adds a
             sketch of Figure 6.2*].

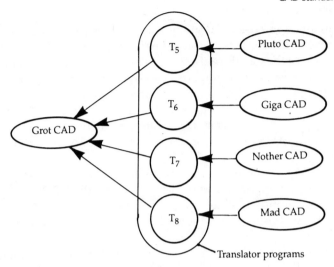

Figure 6.2 *Drawing transfer – in*

DW: Using IGES, we don't need to write any programs at all. The programs all exist. The process of communicating with Coolpoint would look like this: [*Figure 6.3*].

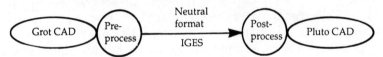

Figure 6.3 *Drawing transfer – IGES*

AS: [*Relieved*] Sounds OK. What's the snag?

DW: There isn't one, apart from the complication and time taken for translation. Oh yes, and IGES doesn't support solid modelling.

AS: But we don't use solid modelling....Do we?

DW: No, we don't. And that's something else I would like to talk to you about. We could do with...

AS: Thank you, Daisy; must rush now. I must go for a haircut.

■   ■   ■

Drawings are currently transferred between systems; the major reasons are:

1. Company A has to send drawing information to Company B. This may be because Company A is supplying parts to Company B and Company B needs to enter the details of these parts into its system. Another need for transfer will arise when Company A is buying parts from Company B and wishes to send a specification of the parts required.
2. A company has more than one system and each has a different method of holding data. It is necessary to transfer information from one to another - for instance to transfer draughting information to a part-programming system.
3. A company has more than one draughting system, the systems being from more than one source and so having more than one format. This is reasonably common in large firms. Even though there is more than one system, it is desirable that there should be only one drawing record system.
4. A company has an existing system which it has outgrown and wishes to buy a better one. It is desirable that drawings should be transferred from one system to another without re-drawing which is time-consuming and error-prone.

Methods of Transfer

1. We can read information directly from the drawing by either automatically or manually digitizing it. There are several systems that can automatically read drawings by scanning them. However, this method is not really satisfactory. The accuracy available is not normally sufficient, we usually wish to transfer more information than the actual lines drawn and the systems now available are either very costly or unreliable. This method may be adequate for schematics, however.
2. We can write a special program to translate from one format to another. This is the most efficient method in terms of translation time, but the problems involved have been outlined in the previous scene. It is no simple matter to write a translator program and the effort required in writing, testing and de-bugging will be considerable. Because of the lack of generality, it is necessary to write a translator program for each combination of systems involved. There have been more than a hundred draughting systems on the market at any given time during the last few years and it is unlikely that this number will decrease significantly for some time. In order to translate directly between each pair of a hundred systems, close to 10 000 translator programs would be required - this represents a lot of programming time, to say the least. Although direct translation is the cleanest way to translate between two systems, it is not really an option for most firms.
3. We can pre-process the drawing file from the original (or 'native' format) into a standard (or 'neutral') format; then we can post-process the drawing file in the neutral format into a format acceptable to the second

system. This takes up a good deal of computing time, but there is a large saving in the amount of software which needs to be written. For the one hundred systems mentioned, we would only need two hundred translators - a considerable saving. A widely-used standard neutral format is IGES; pre- and post-processors to and from IGES are provided by most major systems.

Drawing Information

The file describing a drawing contains various types of information. Since the drawing is a major source of product information, the file normally includes details of more than the shape of the component. Drawings usually contain, for instance, manufacturing information, parts lists and details of gauging methods. If the drawing has been done on a computer-aided draughting system, then there will also be a need for information about layer usage, symbols and other features. A two-dimensional drawing file might contain the following classes of information:

Geometric Data

This is a description of the shape of the part and is a definition of all the points, lines, arcs and circles of which the shape is made up. Many systems do not permit anything to be drawn which is impossible with ruler and compasses; some allow curves which are of a freer form than arcs of circles and are more like those drawn with French curves, usually called 'splines'. Each point, line, arc or circle (collectively called 'entities') will also have information about the line thickness, line colour and line font in which it is drawn. An example might be:

```
CIRCLE
CENTRE: X=120, Y=140
RADIUS: R=40
LINE FONT: CHAIN-DOTTED
LINE THICKNESS: .5
LINE COLOUR: 4
```

Geometric entities would be more likely to be coded and not so easily readable by humans as this example; some systems (AutoCAD is an example) allow alternative formats - a compact one, for economy of disk storage, and an easily readable one, for ease of manipulation by user-written programs.

Text

Drawings almost always contain text, which is used for various purposes - general notes, manufacturing information, explanation of things which are not obvious from the drawn geometry and so on. Text may also be stored in the file in a form similar to:

```
TEXT
DO NOT SCALE
POSITION OF BOTTOM LEFT: X=100, Y=500
ANGLE: A=0
HEIGHT: H=3
LINE THICKNESS: .3
```

Again, it is unlikely that a real system would have such a readable format. Also, many systems would provide more extensive text facilities than this - examples are choice of different text styles and width/height proportions.

Dimensions

The normal dimensions shown on drawings are a mixture of geometric entities (lines and arcs) and text. They may be described in much the same way as the examples above, but there is a difficulty. All realistic systems permit automatic dimensioning, which means that a dimension must be tied in some way to the geometry that it is defining. If we dimension the side of a rectangle and decide later to change the length of the side, then it is convenient if the dimension changes automatically also. This is known as 'associative' dimensioning and it requires some complication in the file structure - not only the lines, arcs and text making up the physical dimensioning must be shown, but also the dimension must be tied to the geometric features that it defines.

Layers

As we have seen, drawings in computer-aided draughting systems are organized into layers. There is often a Drawing Office standard governing the use of layers - layer 1 might be used for the drawing sheet, layer 6 for hatching and so on. Details of layer utilization are held in the drawing file.

Symbol Libraries

An essential feature of draughting systems is that it should be possible to store and retrieve standard symbols at will; for instance, if we were involved in the drawing of electrical schematics, it would be highly inconvenient if there were not a standard set of electrical symbols that we could use. If we wanted to insert a capacitor into a schematic, for instance, then we would call it up and specify where we wanted it placed and at what angle. Since the symbol is already stored in the system, it would be a waste of file space to hold its drawing details in the drawing file. All that is necessary is the symbol name and details of its position in the drawing. Not only does this save file-space, it also has the advantage that if the company standard should be changed then all that is needed is to change the symbol definition rather than laboriously to modify each drawing on which it occurs. This is another example of the 'space–time' compromise that always exists in computer applications; although we have saved file-space, we have had to pay the penalty of longer processing times.

Connectivity

Some advanced systems provide functional connectivity between drawn objects. An example of this in electrical design is where a schematic may be input to a system which will do performance calculations on it; lines drawn between components,in this case, are not just geometrical entities but also represent physical connections. The drawing file must show that these lines are to be dealt with in a different way to the lines that represent a transistor.

Attributes

Most of the larger commercial systems allow 'attributes' to be attached to drawings. In manual draughting, the paper drawing is an important source of product information though there is always other information scattered around the different departments of a firm. If we wish to use a computerized system really efficiently then, as we have previously discussed, it is necessary to have a centralized product database. Part of this database will be the geometrical description of the product; other information will be carried in the product description file. This is known as 'attribute data'. Some systems (PAFEC Dogs, for instance) permit 'property' data to be used to carry out running calculations as the drawing is proceeding - in this way, schedules of weights, lengths and costs can be constructed automatically.

Most large systems carry all these types of data in the drawing file and it should be easy to see why the definition of a neutral format is no straightforward matter. Every system has its own format for its own set of facilities and each system vendor claims that his or her own particular 'goodies' are better than those of the opposition. If perfect transfer were possible between any two systems, it would imply that there was little to choose between them.

A Neutral Format - IGES

The most widely used neutral format is IGES. An IGES file is in five separate parts. Briefly, these are:

START SECTION   This contains any useful documentary information which may be read by the user. There might, for instance, be a description of the drawing facilities of the source system. This section is set up manually by the originator of the file.

GLOBAL SECTION   This contains information which may be used by the post-processor to set up the drawing correctly. It includes file-name, units, scale, date and creation time and the author's name. Again, this is set up by the originator.

DIRECTORY SECTION   This section is generated automatically by the IGES pre-processor. It consists of a list of the entities that make up the drawing with pointers to their positional details which are contained in the next section. Each type of entity is allocated a code and the full list of possible entities is quite extensive (although most, if not all, available pre-processors only contain a few of the possibilities). Not only are geometric entities held here, there are also annotation entities such as dimensions and notes, and structure entities which permit advanced facilities to be used.

PARAMETER DATA SECTION   For each entity entry in the directory section there is a list of data in this section. For geometric entities, such as lines and arcs, there are coordinates. For annotation entities, such as notes and dimensions, there are details such as the text to be drawn.

TERMINATE SECTION   This signals the end of the IGES file.

Limitations of IGES

The aims of the IGES format are to permit the interchange of data accurately, easily and efficiently. In practice, it does not achieve any of these aims completely.

1. Few, if any, pairs of realistic systems can transfer all possible drawings accurately and it is usually necessary to know something about the sending system in order to interpret the IGES error messages and take remedial action. However, it is possible, if there is to be a lot of drawing transfer between centres, for an agreement to be made that the transmitting centre should use only the facilities which are capable of being post-processed by the receiving centre. This solution has its problems.
2. It is certainly very easy to pre-process system drawing data and very little human intervention is needed. The Start and Global sections are set up by the user and are easily understandable. For the most part, the information contained in them may be a defined Drawing Office standard. The post-processing is often not easy, however, and drawing recovery requires a good deal of detailed knowledge of the systems at either end.
3. Whether IGES is an efficient system or not depends on what we mean by 'efficiency'. Because of the generality of their format, IGES files are usually much larger than the 'native' format files. Also, there is an overhead of pre- and post-processing time. In terms of storage requirement and processing time, the use of IGES is highly inefficient. But in modern computing systems it is usually the human resource that is to be conserved and the use of a neutral format aims to do just that. Without it, we would either have to spend a lot of time writing direct conversion programs or laboriously have to re-draw every drawing that we wished to transfer.

IGES is currently the most popular neutral format. It is intended for transfer of normal engineering drawings and is not suitable for any of the more advanced representations, such as three-dimensional solid models. Nor is it suitable for transferring non-geometric product information. It has been in existence since 1979 and is now at level 4.0. Several groups are working on successors to IGES - one is the Product Design Exchange Specification (PDES) which, it is claimed, will solve many of the problems which have been discussed. However, there is a need now for a neutral format and IGES is a temporary, even if imperfect, solution.

MAP and TOP

'Press On Regardless'

Scenario

We return to Palaeolithic. Mrs D. Wheal is talking to Charlie Parker, the Computer Programmer, who also doubles as Chief Operator and Computer

Manager. She seems rather agitated; he is reading the sports page of the 'Daily Planet':

DW: ...and apparently, the drawing files that we sent to Coolpoint cannot be entered into their system.

CP: That's a bit of hard luck. See the Blues pulled it off again on Saturday.

DW: I don't really understand what the problem is. Surely if we translate drawing files to IGES, they will be understandable to any system which has an IGES post-processor. Or am I being naive?

CP: You are, a bit. Even though the information is capable of being understood, it can still happen (and usually does, in practice) that the physical means of communication don't match. For instance, at one time, even tape widths could be different - they seem to have standardized now, fortunately. There are also differences of recording density, different numbers of tracks and so on. The most probable cause of incompatibility in this case is that the codes for letters and numbers are not the same on each machine.

DW: What can we do about it?

CP: Usually, in large installations, there are systems programs available which can convert from one format to another. Perhaps Coolpoint can be talked into getting one. It will be useful for converting tapes from other suppliers.

DW: We'll have a chat to them next week. In the meantime, what about the current batch of drawings? They are keen on getting it as soon as possible.

CP: There is a reliable bureau in town that specializes in such matters. I'll give you their telephone number.

DW: More cost and time, I suppose. I thought all our troubles were over when I discovered IGES. Is this a common snag with computers?

CP: I'm afraid it is. Even though the information to be passed from one machine to another is understandable to both, the physical details of communication may be different. It doesn't just apply to computers, it is the same with any devices which can send or receive data. Just think of the different formats of video tapes. Even though two tapes may contain the same video frames, you have to make sure that it is played on a VTR with the same format as the one that it was recorded on.

DW: Thanks, Charlie. I'd better go and bear the glad tidings to Mr Stickler.

CP: Give him my love...

■   ■   ■

Communication between computer programs is, as they have discovered at Palaeolithic, not straightforward. The difficulties are caused by lack of standardization, in the main. In a factory network, where computers, machine tools, robots, automatic guided vehicles, programmable controllers and other devices must communicate, the difficulties are even more extreme. Some of these are:

(1) the information sent may not be organized in a way that is suitable for the receiving device;
(2) there may be differences between the means of transfer;
(3) there may be differing requirements of timing between sending and receiving.

1. PlutoCAD could not understand what GrotCAD was saying. A solution was to translate one format to another; although this probably reduced the capabilities of each system, communication became possible. It was argued that, for flexibility and cost, a two-stage translation process was preferable. This is a bit like communication in natural languages, where many artificial languages have been invented for a similar purpose. If everyone learned Esperanto, nobody would need to learn any other foreign language. The penalty would probably be that powers of expression were reduced - it would be difficult to use Cockney rhyming slang, for example. The problem does not only occur in communication between programs. In a CAM system, for instance, we need a post-processor to convert the system's format to one acceptable to a machining centre. There are intermediate standard formats available to ease the process.
2. Even if one element of a network is capable of understanding what

another is saying, the means of communication may not be compatible. Anyone who has worked with microcomputers knows the problems involved in connecting non-standard equipment to a system. A program may be perfectly capable of sending data to a printer but if we haven't got the right connector, the information cannot be sent. A worse problem might be that the interfaces were different. If we did not know Italian and wished to give a message to an Italian who could not speak English, we might employ an interpreter. But, if we were eccentric enough only to speak over the telephone, and the Italian was deaf, then, even though the problem of *comprehension* might have been solved by interpretation, we would still be left with a problem of *connectivity*. Palaeolithic have solved the problem of comprehension but they are still left with the problem of connectivity.

3. In an integrated system, a variety of computer programs and devices are connected together and must be able to speak to each other. However, the timing requirements of, say, sending a file from a draughting system to a part-programming system are different from those of sending a command from a controller to a machine tool. In the first, the time lapse between sending and receiving is not very important: a user might become mildly irritated if response time in a computer system was consistently longer than a second or so, but this would not usually be disastrous. If, in contrast, a robot was picking up parts, loading them to a machine tool and then removing the machined parts, it would be vital for the commands to be rapidly and consistently transmitted or else the results *would* be disastrous.

If, in a noisy bar, you ask someone to have a drink and they cannot hear your voice above the juke-box, you can repeat your message until they respond. This would be satisfactory because the message that you wish to send is not vital to either of you (unless, of course, it is getting near to closing time). If, however, you were warning them not to sit on a broken seat, you would probably choose some method of communication other than speech, which was speedy and which would have a guaranteed effect - such as tapping them on the shoulder.

Why MAP and TOP?

It is claimed that General Motors spend more than £4000 million a year on capital equipment. In 1979 they had, in their plants, about 41 000 intelligent devices, including over 20 000 programmable controllers and 3000 robots. According to their predictions, the number will rise to 200 000 programmable devices in the early 1990s. An inquiry revealed that about half their total manufacturing costs were soaked up by the process of connecting these devices together since 85 per cent of them were incompatible with the others. Not only was this extremely costly, it also resulted in delays in using new equipment to maximum capability.

In order to solve this difficult problem, General Motors initiated a plan to produce a factory floor communication standard. They drew on the most suitable existing standards, grouped these together in a logical way and the result is known as the Manufacturing Automation Protocol, or MAP. Normally, standards in computing applications are arrived at in one of two ways. One way is for a committee to design a standard, which is, often after a lot of controversy, adopted by manufacturers. An example is the Graphical Kernel System. Another way is when a standard is adopted industry-wide because of the dominance of one supplier - an example is the screen-addressing protocol used by Tektronix. This is known as a 'de facto' standard. Because of General Motors' vast annual expenditure on equipment, they are in a good position to put pressure on vendors; they buy about three-quarters of the output of the top three USA robot manufacturers, for example. Because of this pressure, MAP has now become a de facto standard and is supported by most of the larger vendors of factory automation equipment.

In parallel with the development of MAP, Boeing Computer Services have produced the Technical Office Protocol or TOP. TOP is specifically designed to enable the connection of office equipment, where timing is not such a crucial matter as on the factory floor. It is important to stress that these are standards to establish *connectivity* between devices; they do nothing to ensure that one system or device will *understand* another.

How MAP and TOP Work

Both MAP and TOP use a bus network which connects together all the elements of an integrated system. Typically, a MAP bus is a co-axial cable linking together all the centres of interest in a factory. Although this cable can pass several messages at the same time, each carried at a different frequency ('broadband' transmission), MAP uses only two channels, one to transmit data from sending centres, one to re-transmit it to receiving centres. These centres are often groups of devices which may also be separately interconnected - the so-called 'islands of automation' which we have met before. It may be sufficient to connect these together by single channels, which is cheaper than broadband transmission. Other channels on the bus may be used for other factory-wide communication applications such as energy management.

Transmission over a MAP network is highly controlled. Stations can send messages only when they have the authority to do so; this is granted by a 'token', a signal that moves along the network visiting each station in a fixed order. A station wishing to send a message must wait until it receives the token, which it holds for a fixed length of time. On transmission, the message passes along one channel until it hits the 'headend' which re-transmits it along the second channel at a different frequency. This frequency can be received by all the stations on the bus and so the message can be picked up by all the stations to which it is addressed. Since only one message is carried on the bus at a time and each transmitter is given a fixed amount of time, the time between sending and receiving can be guaranteed. This is called 'deterministic' communication.

TOP acts in a different way. It operates on only one channel and any station can send a message if the channel is not carrying another. The presence of a carrier signal on the bus indicates that a message is being sent so every station wishing to transmit must 'listen' for the line to become free. When the channel has become unoccupied, the station can send its message and continues to do so for as long as it wishes. The only complication is that two stations might realize that the line is free and start to send simultaneously. In this case, the 'collision' of the messages is sensed and both are stopped. They must wait for another opportunity. A characteristic of this type of communication is that the time between wanting to receive and reception is variable and so it is used where response time is unimportant. It is called 'probabilistic' communication.

MAP and TOP are protocols to support these modes of communication in a standard way. Like other modern communication protocols, they are of layered structure. Both are divided into seven layers, each performing a different function. Layers 1 and 2 (the 'physical' and 'data link' layers) handle the physical transportation of data between nodes. They are responsible for defining the transmission medium, speed and error detection. Intermediate layers 3-6 ('network', 'transport', 'session' and 'presentation' layers) control the routing of data, its reliability and security, re-synchronization in case of failure and standard data encoding. The highest layer, 7 or 'application' layer, provides an interface for specific applications. MAP and TOP are the same at layers 3-5 but differ at layers 1 and 2 because of the differences between a token-passing bus and a collision-detection bus. Of course, these technical details are transparent to the normal user; this is similar to a user of a system needing to know nothing about the programming language in which it is written.

The Graphical Kernel System

Introduction

If managers are to make correct decisions, it is vital that the information on which they base the decisions should be of high quality. Not only must the data be correct and up to date, it should also be easy to interpret. During the 1960s, when batch systems were widely used, it was common for data to be presented to the user in a thick wad of lineprinter output. Even if it was accurate, because of the volume and poor presentation it was often difficult to analyse. It is now accepted that the most natural and powerful way of communicating information between computer and human is by graphical display: graphics is used not only in applications like CADCAM but in business, finance and marketing applications also.

Computer technology is, to a great extent, demand-led and the demand for good quality graphics has resulted in a boom in research and development in graphical equipment. Graphics equipment is no different from any other kind of computer equipment in that there is little effort made to standardize on interfaces and so individual manufacturers provide their own protocols and graphics capabilities. Another problem is that, as is evidenced by the brief treatment in Chapter 4, graphics techniques are mathematical and though the commoner algorithms are now well documented, they are not easily incorporated into programs by the average programmer.

It has been recognized since the early 1970s that there is a need for graphics standards and many attempts have been made to establish them. Although a very good case can be made for the definition of computer standards in general, it is sometimes argued that rigorously imposed standards inhibit technical progress and there is some truth in the argument. From the standpoint of the ordinary user, however, there can be no doubt that a good, widely-adopted standard simplifies the development of systems. In order to be really beneficial, a standard should be *machine independent* so that applications programs can be transported from one computer to another with a minimum of inconvenience, should be *device independent* so that an application program can be used with a variety of input and output hardware without having to be re-written and should be *programmer independent* so that application programs can be written by non-specialist programmers.

Probably the most popular current standard is the Graphical Kernel System (GKS) which was originally developed by DIN, the Standards Institution of West Germany, and since has been adopted by the International Standards Organization and the National Standards Institute of the USA.

Features of GKS

GKS provides a set of procedures for performing simple graphical operations in two dimensions. The four main graphical routines or *primitives* are

POLYLINE (N, XPTS, YPTS)
POLYMARKER (N, XPTS, YPTS)
FILLAREA (N, XPTS, YPTS)
TEXT (X, Y, string)

POLYLINE draws a set of N-1 continuous lines whose endpoints are given by the coordinate values stored in the arrays XPTS and YPTS each of length N. The call POLYLINE (5, XA, YA) would produce the display shown in Figure 6.4 if the arrays were

XA(1)	=	10:	YA(1)	=	10;
XA(2)	=	20:	YA(2)	=	10;
XA(3)	=	20;	YA(3)	=	30;
XA(4)	=	10;	YA(4)	=	30;
XA(5)	=	10;	YA(5)	=	10.

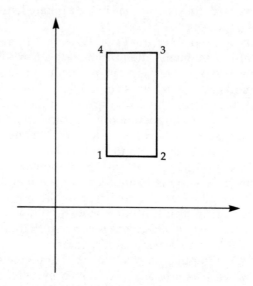

Figure 6.4   *GKS primitive POLYLINE*

Associated with POLYLINE is a function, SET POLYLINE INDEX (N), which sets the line font: the line type (whether solid, dashed, dotted etc), the line width and the line colour. These may be workstation-specific to take advantage of the hardwired functions now available in most displays.

POLYMARKER places symbols or markers at each of the N points defined by the coordinate values stored in the arrays XPTS and YPTS. With the arrays XA and YA defined previously, the call POLYMARKER (4, XA, YA) might produce Figure 6.5. Different markers may be written by previous use of the call SET POLYMARKER INDEX (N) where N is an integer parameter which defines the marker form, size and colour.

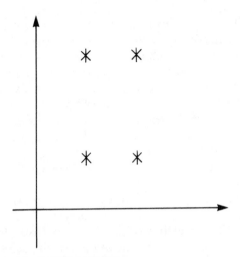

Figure 6.5 *GKS primitive POLYMARKER*

FILLAREA fills in the space within the points given by the arrays XPTS and YPTS with some fill pattern. If the points do not define the vertices of a closed area, the last point is automatically joined to the first so as to close the boundary. Different fill patterns can be used by a previous use of the call SET FILL AREA INDEX (N).

TEXT prints the string of characters given as a parameter at a position defined by the two coordinates X and Y. The call SET TEXT INDEX(N) selects the type font from a list of alternatives such as Italic and Roman but, since text is more complex than drawing lines, planting markers or hatching closed areas, it has attributes other than its geometric form and colour. These attributes may be set by standard routines and include text height, text angle, text direction (left, right, up or down) and text justification.

Drawing a picture on a graphics display is very much like plotting one manually on graph paper. We have a figure to be plotted which is dimensioned in some way. In order to plot it, we have first to scale it to some convenient size and convert the system of dimensions to Cartesian coordinates. Often, we do not use all the paper, but just part of it; we also might display more than one drawing on the same sheet. Sometimes, we only wish to draw part of the picture. These requirements are all formalized in GKS. The item to be displayed is in *user coordinates*, possibly polars or any other representation; in order to display it, since screen addressing is in Cartesian coordinates, we must convert the user coordinates to Cartesian form - the *world coordinates*. These must be converted to *device coordinates* for plotting to take place. But, as has been discussed , different displays have different addressing systems, and it is a requirement that GKS should be device-independent.

This difficulty is overcome by using *normalized device coordinates* with an imaginary or virtual display which is addressable in the range 0–1 in both horizontal and vertical directions. Normalized device coordinates can be converted to device coordinates by a separate program and need not concern the average user. The programmer can define a rectangular portion of the picture, or *window,* in world coordinates. A corresponding rectangular section of the display can be defined in normalized device coordinates; in this *viewport* can be displayed the contents of the window. GKS supports multiple windows and viewports so that complex multi-view displays can be defined. Drawing may be confined to the viewport boundary by setting a clipping indicator.

Drawings may be generalized by grouping primitives and other commands into *segments.* A segment may be dealt with as a single graphical object and transformed, using the matrix techniques described earlier. Thus, segments can be moved about the screen, rotated about fixed centres and scaled.

The use of normalized device coordinates is an example of the way in which GKS supports device-independence of graphical output. It is easy to attain independence of output, since output devices, whether display screens or plotters, may be treated in much the same way. This is not true of input devices: the variety of their types and functions has been already pointed out in Chapter 4. GKS groups the available devices into several logical categories and deals with them separately. These categories have been mentioned before in Chapter 4. Once the common features have been grouped, it is possible to make the input routines independent of real devices. GKS is also language-independent. In order to achieve this, the defined subroutine names and parameters may be given equivalents in the particular language being used. This is called *language binding* and GKS may be bound to common languages such as FORTRAN 77 and Pascal.

There are two standards associated with GKS. These are GKS-3D and PHIGS (Programmer's Hierarchical Interactive Graphical Standard). GKS-3D extends the capabilities of GKS to include such features as three-dimensional primitives and hidden surface removal. It is downward compatible with GKS, so that existing programs need not be changed if an installation moves from the two-dimensional standard to GKS-3D. PHIGS can work in association with GKS-3D and handles graphical data structures.

Graphical standards, like other computer standards, must be in a constant state of review. At no time can a standards committee state that the standard they have produced is definitive for ever, because new devices are constantly being developed and nobody can foresee with any certainty what the next development will involve. However, most system designers would agree that an imperfect standard is better than no standard at all.

Problems

1. Investigate the drawing files that your system produces. Write a report explaining the format to an ordinary user.

2. Three graphics standards are

   (a)   GKS
   (b)   The CORE
   (c)   GINO-F

These have different historical backgrounds. Write a brief account of the development of each.

3. 'Standards are counterproductive because they inhibit system development.' Do you agree with this opinion? Give reasons.

Postscript

AN OPEN LETTER FROM NED LUDD

Biographical Note

Ned Ludd (or Lud) was a possibly mythical early machine breaker whose name was adopted by the Luddites (1812–1818). This was a band of workers who destroyed machinery in the North and Midlands of England because they believed that it caused unemployment. At the height of the movement, over 10 000 troops were deployed against the Luddites.

Dear Reader,

You have been told in this book that CAD is probably the best thing to happen to engineering since the slide-rule, but I hope that you have not been convinced. I consider that CAD (or anything else starting with the letters 'CA' for that matter) is a waste of time and money for manufacturing firms, and a bad influence on design, that it leads to poorer products and has a disastrous effect on the working conditions of engineering staff. Let me elaborate.

Effect on a Firm

It cannot be denied that many systems promise more than they deliver. In the early days of CAD, it was commonplace for vendors to make wild claims about the productivity increases that could be achieved using their systems - a factor of ten was then quite a modest claim. It is noticeable, now that systems are cheaper and more widely used, that they have moderated their claims. Vendors who once hinted that, if you bought their system, you could do away with most of your design staff, now state that there will not necessarily be any increase in drawing throughput, but that there will be all sorts of intangible benefits. Can we trust the sellers of systems? You might answer that you know of many firms who are well satisfied with their system and the benefits it brings. Can we trust them either? No one who has been involved in the procurement of a large, costly and disruptive system is likely to admit that it is no use whatsoever.

Once a firm has decided that it is going to implement a CAD system, it is starting a process that cannot easily be reversed and that will result in a continual drain on its resources. Systems and hardware are constantly being 'improved' and new CAD products are frequently announced. The life of an existing piece of equipment or software is steadily reducing and so the firm has to keep spending money - it is almost impossible to go back because of the time invested in the system and the organizational upheaval that would be needed. It is often suggested that a firm considering investment in a CAD system should be cautious and enter the field at a modest level before going whole-heartedly for a full-scale system. But a small inferior system cannot adequately test the suitability of CAD for their needs; on the the other hand, to purchase a full-blown system initially may lead to financial disaster - computer-aided bankruptcy.

Integrated systems, we are told, ensure that all the information held by a firm is up to date and consistent. But a fallible human being must enter most of that information to the computer and there is an old computer saying 'garbage in, garbage out'. Although integrated systems do ensure *consistency* of information, we have not advanced much in ensuring its *quality*.

Another problem with centralized information is that it is vulnerable to industrial espionage. It is difficult, if not impossible, to prevent a determined and skilled person from getting access to privileged information. If a hacker can get access to central government files, as has happened, what chance has a small firm of keeping its data secure? In an uncomputerized system, the data is dispersed in various forms around the departments of the company; although managers cannot easily and quickly get access to information, neither can electronic interlopers. Not only are centralized systems susceptible to unauthorized access to data, there is also a risk of all the data being lost because of a natural disaster, such as a fire. You can take all the precautions you want, take security copies every hour if you wish, but no system can be proof against loss of data. The highly improbable is still possible.

Effect on Creativity

In order to perform a task by computer, it is necessary to define it precisely so that it may be reduced to a set of simple well-defined operations. If computer systems are to be used to help designers in the tasks that they do - and perhaps replace designers in at least some of them - it is essential that the design process should be analysed and broken into its component parts. But designing a component is not like making a cup of instant coffee. There is no universally agreed method of working. There is considerable disagreement about the nature of the design process and it is generally accepted that designers appear to work in many different ways. If it were possible to derive a standard model for design, then it would certainly not suit all designers; possibly, it would suit very few. If we attempt to force designers to work in a way that is unnatural to them, then we run the risk of cramping their creativity. There is no guarantee that the best, most inventive designers, who often work in a highly individualistic way, will be able to generate the same ideas if they are forced to work with a computer system.

An example is in the use of solid modellers. We are told that their use will enable the designer to create a geometric model which may be displayed realistically, rotated and viewed from various aspects, then modified to the designer's taste. This is just not possible on current systems. The model must be precisely defined dimensionally, must be processed by the system (which sometimes takes hours for a complex model) and cannot be modified without all the processing being repeated. I know of no designer who works like this. It is usual to develop a design from rough sketches, evolving the form by a system of trial and error and only bothering about the physical sizes at a late stage in the procedure. This method of working is impossible on current systems and if designers are forced to work in such an alien way, then their creativity will certainly suffer.

Another problem is that the system often imposes itself on the design - I notice that it was pointed out earlier that three-dimensional modellers can only model a fraction of the possible shapes that a designer might require. Design is a difficult business already, without adding more obstacles.

A further danger is that some talented designers will channel their creativity into using computers for no other reason than that they are interesting machines and challenge the ingenuity of the user. We have few enough skilled designers already, without some of them becoming computer 'hackers'. It is commonly observed that some people are fascinated by computers, and we have all heard tales about students who spend almost all their waking hours at a terminal, having sandwiches brought to them, and writing programs that are of no interest to anyone apart from themselves. Stories about computer hackers getting divorced because they relate better to a machine than to their husbands or wives are widely told.

If a designer were to prefer communicating with a machine rather than with human beings, then it would be disastrous. It is recognized that 'synergy' is useful in the production of new concepts - human beings are at their most imaginative when they can try out their ideas on other human beings and receive stimuli from them. Even worse, we are told that with the latest advances in telecommunications, a designer will not even have to leave home; the system will be available over the telephone. This is in line with the modern trend towards isolationism - once public transport was the preferred means of transport, now we like to travel in our own private car; once the cinemas were filled, but now everyone has their own private television. What sort of design stimulus can we get from a machine?

It is stated, by people who ought to know better, that CADCAM is a good thing because manufacturing decisions can be made at the design stage. Ask most production engineers what they think of that. Most designers also would admit that they feel most comfortable when they have only the function of the component to bother about and that the best designs come about by collaboration between them and production specialists.

Don't think that these bad effects concern only the application of computers to the initial inventive stage of the design cycle. I would argue that there is a good deal of creativity involved in all stages of design. Detail design is an example. The success of a design is highly dependent on good detail design, which has a high creative content.

Effect on the Product

Computer systems attempt to produce an optimum design. But consider two firms using the same system to produce designs for, say, automobile tyres. Surely they will both finish up with broadly the same design of tyre. How can either claim that their tyre is better than the other, if they are both obviously the same? They will probably be forced continually to re-design and add extreme styling differences. The public will, of course, suffer - either their choice will be reduced or they will have to pay for unnecessary product development. This effect can already be seen in automobile design; up to a few years ago, it was possible to recognize a model of car easily - there was such a thing as a 'house style'. Nowadays, it is difficult: the popular car is a plain box built around four passengers.

Conversely, using computers may produce the opposite effect, and instead of resulting in excessive re-design, CAD may stifle progress. An example is where a complex range of parts has been parameterized, so that drawings can be produced without draughting. Because of the high overhead of time taken to produce a parameterized version, a firm might hesitate to modify the range even though the modification would result in a cost-saving or in improved performance.

If computers are not used critically, they can lead to a deterioration in product design. You might claim that you are using good quality, well-established software. The problem here is that it is impossible to test thoroughly programs of any realistic size. It has been said that testing can prove the presence of bugs, but not their absence. Software associated with the Space Shuttle project is said to have taken four hundred programmers four years to write; it was still liberally bestrewn with errors. CAD programs are certainly no better. If you look at the small print on a software contract you will probably see a disclaimer clause, freeing the vendor from liability for any losses resulting from bugs in the system. This lack of confidence is not shared by some users, who have a misplaced trust in the answers produced by their system: if these answers are wrong, they may pass undetected from department to department.

Effect on People

I have argued that CAD has a bad effect on the firm, the product and on the creativity of the designer, but it is in its effects on the working conditions of the engineer that its effects are worst.

CAD creates a split between those who use systems confidently and those who do not - it results in the formation of a sort of computer elite. You may think that this is no bad thing, but it is customary for engineers to complain about the decision-making in a firm being made by the accountants, who have no commitment to the product beyond that defined by the balance sheet. The use of CAD means that decision-making is in the hands of computer experts. Is this any better?

Draughting was once a highly respected and skilled job; there was a personal draughting style which took many years to achieve and in which draughtspersons took a good deal of pride. It was possible to look at an engineering drawing and to be able to tell who had drawn it from the quality of the communication: the printing, the layout and the clarity of the drawing. CAD minimizes draughting skill. The expertise now is in the efficient use of a computer system and not so much attention is paid to the product.

Drawing office personnel now often have to work shifts so as to utilize the equipment fully. And the work is more arduous; it is very fatiguing to work at a terminal for long periods. Fatigue is not the worst effect, since there are indications that prolonged exposure to computer terminals is hazardous to health.

I have no doubt that CAD reduces the quality of life of the engineering designer. But the most pernicious result of computer systems in general is that they reduce the number of staff. It is invariably claimed by firms that introduction of CAD and the consequent increase in drawing throughput will not cause redundancy but that the shorter design lead-times will contribute to a more competitive product and a higher share of the market. It is difficult to believe this. It seems that nearly all manufacturing firms will be involved in CADCAM by the end of the century. Are they all going to have a higher share of the market? No, CAD systems cause unemployment. That is what they are for. And since the level of de-skilling will move continually higher in firms, the very people who are pushing CAD today may well find that their jobs are in jeopardy in the future.

Just remember, CAD IS BAD.

Yours sincerely,
Ned Ludd

Problem

The letter from Ned Ludd contains a mixture of half-truth, illogic and inaccuracy. You are invited to write a letter refuting the main points that he has made.

CAD Systems index